Nuclear Powe
Siting and Safety

Nuclear Power: Siting and Safety

Stan Openshaw
Department of Geography
Newcastle University

Routledge & Kegan Paul
London, Boston and Henley

First published in 1986
by Routledge & Kegan Paul plc

14 Leicester Square, London WC2H 7PH, England

9 Park Street, Boston, Mass. 02108, USA

Broadway House, Newtown Road,
Henley on Thames, Oxon RG9 1EN, England

Set in Times 10 on 11pt
by Columns of Reading
and printed in Great Britain
by St Edmundsbury Press Co Ltd,
Bury St Edmunds, Suffolk

Library of Congress Cataloging in Publication Data

Openshaw, Stan.

Nuclear power.

Bibliography: p.
Includes index.
1. Nuclear power plants – Location. 2. Nuclear
power plants – Safety measures. I. Title.
TK9153.064 1986   363.1'79   85-8202

British Library CIP Data also available

ISBN 0-7102-0183 4 (c)
     0-7102-0651 8 (pbk)

Contents

Figures

FIGURES

Preface

It is not often that geography has an opportunity to contribute something which is obviously geographical as well as relevant to public policy and decision-making. The siting of nuclear power facilities provides one of these rare opportunities. The siting process has in many countries been neglected as an area of study, whilst power utilities only view location as a variable that influences the economics of operation and construction. However, the locational component is also a very important variable that will largely determine the consequences of any reactor accidents, should such improbable but not impossible events ever occur. In recent years there has been a tendency to rely only on engineered safety measures even though the possibility of reactor accidents cannot be eliminated by these means. Yet the importance attached to the residual risks of nuclear power varies tremendously between the UK and the US. It seems that UK safety agencies are far more confident than those in the US. The British public is told that harmful reactor accidents will not happen in the UK because the reactor operators have a statutory duty to ensure that they do not. It would appear that the degrees of uncertainty, perhaps considerable, inherent in such confidence are ignored. Will British-made reactors always perform within their design specifications? Will the safety sysems always work sufficiently well to preclude the possibility of accidents? Can operator error really be eliminated by training programmes, by staff selection procedures, and by the use of microprocessors? Will nothing unexpected ever happen? Will terrorists always be reluctant to attack nuclear plant? Prudent politicians and publicly responsive power utilities would do well to take heed and plan to exploit the population geography of their countries fully to ensure that, should the unlikely ever happen then the subsequent public and political ramifications will not close down most of their facilities on the grounds that far too many people are at risk. The additional marginal costs of nuclear electricity from remote sites would be a small insurance premium to pay. It would also do much to increase the acceptability of nuclear power by an increasingly knowledgeable and sceptical public. At a time when many countries are looking forward to an all-nuclear future, or at least a major expansion of the nuclear electricity share, the arguments in favour of a geographically realistic and safety conscious siting strategy need to be elaborated.

Chapter 1 provides a brief look at global patterns of energy development and attempts to identify the likely long term trends. It should be noted that the UK is no longer a major nuclear powered country. In 1982 the 6.9GW(e) 6,900MW(e) or megawatts of nuclear capacity, equivalent to about 6 large PWRs is fairly small. The US had 57.2GW(e), France 21.8, Japan 16.6, West Germany 9.8, Sweden 6.4, and Canada 5.6. It is clear also that nuclear power is still in an early stage of development with large expansions expected. By 2000 AD, the US may have 110-140GW(e), France 54 to 80, Japan 37-60, West Germany 20 to 29, and the UK 14 to 20. It is the need for additional sites and the trade-offs between safety, environment, and economics which makes the next decade – when most of the major long term siting decisions will be made – so interesting and important.

Chapter 2 describes in some detail the power station planning and siting policies used by the UK Central Electricity Generating Board (CEGB). This power utility is fairly unique as an independent state controlled monopoly with long experience in the siting of nuclear plant.

Chapters 3 and 4 outline the various siting strategies used in the UK and discusses the use of automated site search methods to identify the locations of most, if not all, feasible locations in Britain for nuclear power developments. Chapter 5 focuses on three case studies and critically examines the logic and basis for the selection of these nuclear power sites.

Chapter 6 investigates US siting practices and siting criteria. There is a fairly detailed discussion of the very different regulatory approach that has been adopted and of the underlying rationale.

Chapter 7 looks at some of the technical problems with demographic siting criteria and performs a detailed comparative study of the reactor sites selected in the UK and US.

Chapter 8 focuses on the need to identify optimally safe and maximally acceptable sites as being the best way to ensure the long term continuous acceptance of nuclear power as an energy source. Some indication of where these optimal sites may be found is given.

Finally, Chapter 9 offers an overview of the arguments made in the book and offers some suggestions for the future.

Stan Openshaw
May 1984

Acknowledgments

The help, advice, and or encouragement of the following people are gratefully acknowledged: Neil Wrigley for suggesting the book in the first place; John Fernie for providing encouragement and material assistance with the US siting aspects; Martin Charlton for computing advice; the night operators of NUMAC who made the computer runs happen; David Rhind who provided the 1971 population data and the ESRC Data Archive for the 1981 population data, John Knipe for some of the figures, and members of the CEGB who patiently tried to convince the author about the errors of his ways.

1 Nuclear safety: some reasons why siting is important

Introduction

Before looking at the safety arguments which are crucial to the development and public acceptability of nuclear power, it is useful to review briefly some of the inevitable technical jargon. In the UK the first nuclear reactors for power production were carbon dioxide cooled, graphite moderated, natural uranium reactors; the so-called MAGNOX series, named after the alloy used for their fuel cans. The next generation of reactors used slightly enriched uranium, stainless steel fuel elements, and carbon dioxide cooling; the so-called Advanced Gas Cooled reactors (AGR). A few other species that are occasionally mentioned are: the high temperature reactor (HTR) a helium cooled, ceramic fuel version of the AGR; the steam generating heavy water reactor (SGHWR), a UK version of the Canadian CANDU design; the fast breeder reactor (FBR) which is cooled by a liquid metal (sodium) and will breed plutonium from depleted uranium – this is viewed as essential for the future of nuclear power because of this ability to make best use of the available uranium supplies; and the pressurised water reactor (PWR) which is cooled by water under extreme pressure – this reactor seems to be the world preferred thermal reactor until such things are totally replaced by FBRs. By contrast, in the US most of the reactors are of two basic types: boiling water reactors (BWR) and pressurised water reactors (PWR); of which no examples yet exist in the UK although the intention is to build only PWRs in the future. The US family of thermal reactors are thought to offer inherently better economics than the UK's MAGNOX and AGR systems, but they also seem to have poorer safety characteristics under accident or operator error conditions. The AGR can probably be badly abused for several hours before something nasty starts to happen. With the PWR, as Three Mile Island demonstrated so well, the critical response time is counted in minutes. Incidentally, with the FBR the critical time is counted in seconds.

It is now abundantly clear that the future source of electricity in many countries is seen as being nuclear, rather than fossil stations. The dependency on nuclear generation will gradually increase until for many advanced countries it will become the major, and then soon after the only power source, some time in the second or third decade of the next century. However, whilst this end-state may seem a long time in the future (about 40 years) in fact the critical decisions will be made fairly soon. In the UK the Sizewell Public Inquiry will eventually decide the future reactor type; it will almost certainly be a PWR; and it is expected that this decision will be followed by a lengthy PWR building programme. There is a sense of urgency here to ensure that the nuclear commitment becomes so great as to be irreversible. In the UK this state has not yet been reached. The much-heralded nuclear contribution of between 12 and 17 per cent at present could be dispensed with without any noticeable effects. The sense of urgency comes from the need to secure the nuclear commitment before the problems of decommissioning the old MAGNOX stations occur; at present their lives are all being progressively extended, whilst in the US very few of the first generation of nuclear stations are still in operation. It is also urgent before public concern about reprocessing and high level waste storage become so great as to provide a real threat.

The most important decisions will concern the choice of sites. These are of immense significance because of their longevity. Conventional industrial sites may last 20 or 50, even 100 years, but then they can be restored. With nuclear power stations the lifetimes will probably run into hundreds of years because of the virtual impossibility of restoring old nuclear sites to their pre-nuclear condition. It is cheaper and far safer not to do anything other than entomb old plant and re-develop the site with a new generation of reactors. Complete dismantlement of many UK civilian nuclear power stations may be impossible or at least pose major public health problems, and may be economically infeasible on timescales of less than a few hundred years. Similar considerations apply to reprocessing plants; for example, it is exceptionally doubtful whether the famous Sellafield site could be restored on any timescale of less than a few thousand years; if at all.

The future of nuclear power will also bring with it the need to develop additional nuclear infrastructure connected with the nuclear fuel cycle – particularly, more reprocessing plants, high and low level waste repositories, fuel fabrication factories. The siting decisions made here will have as their justification the earlier decisions about the location of nuclear plant. So in many ways the nuclear power station siting decisions will justify others. The siting issue is important because of the long historical timescales that are involved and the need to ensure that sites being selected now will remain acceptable and minimise public health risks both now and throughout the twenty-first and twenty-second centuries, when technologies and public perceptions about radiation dangers might be very different from those of today. Whilst it is genuinely difficult to anticipate what might be regarded as publicly acceptable in 50 years time let alone 250, nevertheless it would be exceptionally myopic to only assume that people of the future will have identical perceptions to those currently held by the nuclear industry. It could be argued that a safer strategy would be to assume that a more aware public would perceive nuclear safety as being an inverse function of their distance from the nearest nuclear plant. This attitude already exists and is manifest in the NIMBY (not in my back yard) phenomenon. As more people become aware of the potential worst-case dangers of nuclear power, perhaps stimulated by the occasional accident somewhere in the world, so this state of mind will probably greatly increase. Of course it is, at present, very easy for national interest arguments to totally defeat such parochial attitudes. However, as more people adopt a similar view so the question increasingly becomes one of whether or not national industries, and indeed commercial firms, have any right to make these sorts of decisions. Once nuclear decision-making goes democratic, then the siting issue will become of immense importance. The basis used to justify locations will fundamentally change from one of secret vested interests (i.e. profit or engineering) to matters of public importance (i.e. health and safety).

In the short term, the siting aspects of nuclear power facilities are only going to attract public and political interest and attention if there are good grounds for believing that existing siting practices are less than satisfactory, that there will be a demand for new nuclear power stations in the future, and that public safety can be influenced by locational policies. This chapter demonstrates that siting is important as an additional safety measure by providing a broad appraisal of the background relating to the likely future development of nuclear power and a critical commentary on certain aspects of the safety debate. The story line is mainly based on events in the UK but the interpretation that is offered seems to be more generally applicable. A picture is given of an asymptotically increasing nuclear future with important siting decisions being made at a time when important safety and environmental questions have not been fully or honestly answered by those responsible for public safety. Given the varying levels of uncertainties that are involved, the argument is made that, instead of a blind total faith in the complete infallibility of engineered safety measures, a more responsible approach would be to use the location variable as an additional, reactor independent, augmentative safety measure. This is not to argue that remote siting or geographical isolation can make an unsafe reactor safe, but to emphasise that the residual risks from the operation of potentially safe reactors, and more especially the public perception of the risks of operation, can be greatly reduced by using geographical isolation as an additional safety factor. This basic theme is recurrent throughout the book and it is viewed as the principal way in which the geography of reactor siting can materially improve the public safety aspects of nuclear power in all countries with a nuclear power programme.

Global energy scenarios

Before the discovery of fission those who speculated about the future of the world had to consider what to do when the fossil fuels ran out. It was widely expected that solar energy, biomass, wind power, wave and oceanic thermal gradients, and geothermal energy might offer useful alternative sources of energy. In fact a fission-free world was the only world until fission was discovered in 1938 and although inexhaustible, cheap, and abundant sources of energy would be needed to avoid a future Malthusian catastrophe (confidently predicted for some time around 2050 AD) it was not until after World War II that a practical alternative appeared. As Weinberg (1977) points out it was perhaps fortunate for the future of mankind that God happened to legislate that the number of neutrons produced for each neutron absorbed was greater than unity; for plutonium 239 it is a factor of 2.8 for fast neutrons. This law of nature made nuclear power inevitable once it had been discovered and once the engineering problems had been solved.

There is today a firm, world-wide belief that if mankind is to avoid a serious energy supply problem sometime in the future there has to be a substantial nuclear contribution (Foley and van Buren, 1978). Proven recoverable oil supplies will probably only last another 40 years or so. Whilst coal may last for over 300 years the prospect of a massive increase of carbon dioxide concentration with possible global climatic effects might well end the fossil fuel era sooner than might be expected simply from coal depletion rates. Yet the real potential contribution of nuclear power only becomes visible when a time horizon beyond 2000 AD is considered and a large number of fast breeders are in operation.

In the 1970s, something like 72 per cent of the world's population consumed less than 2kW per capita; more than 80 countries consumed less than 0.2kW; and about 22 per cent between 2 and 7kW. The difference between the highest and lowest consumption rates was a factor of 50 (Charpentier, 1976). If a simple assumption is made that in the long term the low consumption countries move to a 2kW rate, and at the same time the world's population continues to grow, then it might be thought that the world's future energy demand might increase from 7.5TW (1TW is about 1 billion tonnes of coal equivalent) in 1971 to between 20 and 70TW in the long term. Hafele (1977) considers that 50TW might be a more realistic target for a population of 12 billion people in 2100 AD and between 11 and 24TW for a 2000 AD population of 6 billion. These forecasts may now be regarded as on the high side but doubtlessly values of these magnitudes are realistic if on a slightly more extended time scale.

In meeting these targets nuclear power is only one possibility although at present it is the only option that is likely to be able to cope. However, the real value of nuclear power will only be realised when the principle of breeding is used. A thermal reactor – for example, a PWR or AGR – can burn rather less than 1 per cent of the available uranium. This makes it economic only to use the very rich uranium ores which have small reserves. Some experts believe that thermal reactors will be constrained by uranium production to a maximum of 1200GW (1GW is 1000MW, 1MW is 1000kW) or about 1200 PWRs. It is expected that much of this capacity could be installed by about 2000 AD so that uranium-fuelled reactors are unlikely to make a large contribution to energy growth after the first decade or few decades of the next century. A fast breeder reactor is designed so as to breed plutonium efficiently and so, it could in theory, burn all the uranium. By using fast breeder reactors it is possible to increase the usage of natural uranium atoms by a factor of 60 to 80. This would have a tremendous impact on available energy supplies. For example, the UK already possesses a stock of depleted uranium which, if it is burnt in fast reactors would have the energy equivalent of the entire coal reserves of the country. It

5

also makes it economic to consider very low yield uranium reserves, perhaps even the uranium content of the seas might become accessible. In theory, between 10^7 and 10^8TW might be available through breeder reactors enough for millions of years and dozens of billions of people (Hafele, 1977). A thorium fuel cycle would double this quantity. It should be noted that fusion power, should it ever work, offers an energy supply of a similar magnitude.

To be successful, breeders require a certain level of thermal reactor capacity in order to provide initial fuel loads of fissile material, and also to hike up uranium prices sufficiently to make breeders an economic proposition. Initially, at least, the size of any future breeder programme and the rate at which breeders could be built is dependent on the amount of plutonium produced by thermal reactors, as well as by reactor breeding gain, the time taken to reprocess plutonium, and plutonium losses during reprocessing. These constraints are only relevant for the first few decades of a breeder programme. Possible construction rates for Europe are estimated to be 4 to 8GW per year within 20 years of starting, 8 to 16GW within 30 years – the critical factor being reprocessing times. Indeed as Vaughan and Chermanne (1977) point out: 'New found environmental or political objections to the construction of fuel reprocessing plants would not reduce the amount of plutonium in existence but would make it unavailable to the breeder and so delay the start of the programme by some years' (p. 552).

Breeders are, therefore, currently the only energy source that can be considered feasible on a time scale of a few thousand years as providers of bulk energy. This explains the UK Atomic Energy Authority's sustained determination to develop this option regardless of cost; there are similar parallels in France and the USA. But if fast reactors are the only hope for the future starting in the late 1990s then it is becoming increasingly important to start developing the associated reprocessing plant which could well have lead times of 15 to 20 years, longer than that needed for the reactors themselves. This also implies that critically important siting decisions may either have been made already or will be made fairly soon, in many countries committed to nuclear power. Furthermore, the problems of developing an acceptable reprocessing capability and of persuading public opinion that the nuclear route will not be harmful to their health are also becoming of considerable significance.

Another aspect concerns the need to convert nuclear electricity into other forms of energy – for example, hydrogen – since it is unlikely that electricity will ever account for more than 25 to 40 per cent of the energy end use (it is about 10 per cent at present) (Hafele, Sassin, 1975). This aspect is often overlooked. So, indeed, is the corollary of the quest for fast breeder power. That is simply, the scale needed for an economic unit is currently set by the transmission grid. There is an approximate historical rule of thumb which suggests that supply units for an electricity grid are a fixed fraction (about 10 per cent) of the total electricity system. Lower values sacrifice benefits of scale and larger sizes lead to unacceptably low figures of supply reliability. The size of current electricity grids are about 10GW and the average kWh travels only about 100km to final consumption (Hafele, 1977). Now the fast breeder has to be seen as including the nuclear fuel cycle. Its size will be determined by the size of the associated reprocessing facility. Hafele (1977) argues that if in the future a reprocessing plant of 3000 tonnes per year is practicable, and assuming the entire fuel cycle to be one supply unit for nuclear power, then applying the 10 per cent rule, the standard unit of supply would be 1000GW(e) or 1TW. This by its very size points to a global organisation. The supply grid would be worldwide in the form of a liquid or a gas. Indeed liquid energy carriers are already transported on this scale. The Persian Gulf already ships about 2TW over distances of 10,000km or so.

A fuel cycle that includes reprocessing, waste solidification, waste disposal, and fuel fabrication would allow regional concentration and large-scale energy parks could be set up on a global scale. This would overcome many of the safety problems of nuclear power since really remote sites on a supranational basis could be selected. However, one imagines that political factors related to security of supply arguments would result in national facilities rather than international ones, with far smaller supply units than those suggested by Hafele. The location of the high pressure gas transmission infrastructure might well dictate the future location of such facilities in the same way that the electricity grid virtually determines the location of thermal nuclear power stations. If process heat, domestic heating schemes, and electricity generation were to be combined then urban siting might still be required, were it to be allowed on safety grounds.

Weinberg (1977) has also speculated about the prospects of complete success of a breeder programme. He considers that a 50TW programme would require 30,000 tons of uranium per year from residual uranium, thorium, and seawater as well as $25km^2$ for the burial of high level wastes. None of these factors would preclude such a programme. It is the malfunctioning of the system, particularly the possibilities of accidents and the traffic in

7

plutonium that could cause problems. If the NRC (1975) meltdown rates for light water reactors were to be characteristic of the fast breeders, then there could be as many as one reactor meltdown every 4 years! The total plutonium in such a system would be about 125,000 tons with about 30,000 tons being reprocessed every year or 100 tons per day; this might well cause problems of proliferation. Unless fusion works we seem to be heading in a collective manner towards a future world that ultimately depends on many thousands of nuclear reactors. Even in a world using only 15TW with 11TW coming from reactors, each of 1300MW(e), there would be 5000 such reactors. Such a world would not be a simple one. According to Weinberg (1977):

> It appears . . . then that the future of our enterprise depends somehow on our devising a nuclear energy system (ie reactors and supporting facilities, their siting, and their institutional matrix) that confronts these contingencies – meltdown and proliferation – fully and unflinchingly with the realization that the system we devise must last for a very long time (p. 770).

A fully developed nuclear system will be one which commits only certain pieces of land to radioactive operations, land that would have to be dedicated into perpetuity. The sites should be selected to accommodate the characteristics of nuclear reactors with the co-location of reactors and reprocessing plants being essential in order to keep the traffic in fissile material and radioactivity to a minimum. A small number of large sites will be required and this would make it possible to incorporate additional barriers against diversion than a more decentralised strategy of siting would allow.

The siting implications are quite clear. Again Weinberg put it like this:

> As we begin to deploy breeders we ought to site them . . . in the full realization that we may be committing ourselves to a siting policy that will prevail for an immensely long time. The siting policy I espouse – relatively few numbers of very large sites – may be evolving inevitably. As it becomes more and more difficult to find new sites the existing sites will expand (Weinberg, 1977, p. 777).

The situation in the UK is in line with these world trends. Nuclear power is currently viewed as being very attractive on account of its economics, as being needed in order to preserve coal for other uses, for environmental reasons – particularly the so called acid rain problem – and for political reasons. Additionally, large export markets are perceived to exist in developing countries (Lane *et al.*, 1977). The increased installation of thermal reactors will lead to worldwide competition for fuel and the need for fast reactors in the 1990s. A member of the Central Electricity Generating Board (CEGB) put it like this:

> . . . the United Kingdom cannot isolate itself from the worldwide competition for fuels of all types and we therefore consider that, beginning in the mid-1990's, it will be essential to make use of fast reactors (Mathews, 1977, p. 745).

It is important to note that in terms of the carefully defined semantics of the debate about nuclear power in the UK, this sort of statement in no way implies any CEGB commitment to the fast breeders or that they have any plans for a fast breeder programme. Until there are firm orders for equipment or planning applications are submitted, they will deny it.

The public safety aspects of these trends are simply stated. It is imperative that such sites are most carefully planned, selected with the long-term trends in mind, and deliberately chosen to minimise adverse health effects under both normal and accident conditions. They cannot simply be allowed to emerge in an incremental way explicitly designed to minimise public attention. Likewise you cannot develop a commitment to an all-nuclear future without an explicit national policy objective being defined. The argument can be expressed that many of the existing reactor sites, in most countries, are fundamentally unsuitable for such developments; mainly because of high population densities in the vicinity of the plants. It would be both dishonest and foolish to simply seek to adapt existing sites selected for the first generations of thermal reactors for the next generation of large-scale integrated nuclear power facilities. There should be a fundamental re-think about the nature of the sites needed for future developments; subsequent generations may well be surprised if considerations of short-term convenience and marginal economics rather than public safety were to dominate the site selection process.

Why is siting important?

As Keeney (1980) points out: 'The siting decision is a fundamental component in many policy decisions regarding energy' (p. 6). It is clear that during the next few decades a large number of very expensive, very large, major energy facilities are going to be built. In many countries this will involve a number of nuclear powered stations, radioactive waste storage complexes, reprocessing plant, and energy transmission facilities. Each of these developments poses a whole series of immensely complex and exceptionally important siting problems that deserve the most careful examination because of the potential public health hazards, and the long historical timescales before complete site reclamation will be practicable. The national interest demands that suitable sites be found, but in both Europe and the US there is no really effective siting policy for identifying the 'best' sites (Meier, 1975; OECD, 1979) or even any adequate definition of how optimality may be measured and the weight to be given to the various conflicting variables. The potential magnitude of the problem is immense and the process of finding sites is of great significance, even if at any one moment of time the slow incremental nature of the changes serves to hide both the likely end result and the full scale of the changes that are occurring.

The siting problem is becoming a very urgent one because of the long lead times that are involved. It is estimated that, if all goes well, the development of a new greenfield nuclear power station site requires between 12 and 20 years before there is any contribution to energy supplies and about 10 years for an existing site. Sir Jack Rampton, Permanent Under-Secretary of State at the UK Department of Energy put it this way:

> So most of the decisions that will affect our energy supplies for the next decade have already been taken and if one is interested in the turn of the century there is not much time in which to decide what to do (Department of Energy, 1979, p. 2).

The planning problems are immense and important decisions are being made under conditions of great uncertainty. These decisions are based on forecasts for the early years of the twenty-first century – about which all we know is that the forecasts will probably be wrong. Long lead times often result in panic decisions and cause understandable anxiety. The need to ensure continuity of supplies is a major factor, but in many countries there is a growing risk that a considerable proportion of current plant will exceed their original design life within the lead times needed for new development. One can speculate that security of supplies is about to become a major problem. The more procrastination and enforced delay there is now, the worse the potential future problems become. However, having said that, it is not clear that the *only* appropriate response is a large new building programme tied to uncertain forecasts of both supply and demand in the early years of the next century. At risk is not only the security of supplies but also the tremendous sums of public and private money that could be wasted should the wrong decisions be made with an irreversible commitment to a publicly unacceptable form of power generation.

Siting is an important factor in the great future energy gamble because it can significantly delay major investment decisions; it can influence marginal costs; and by affecting public safety it may, in turn, affect the acceptability of the decisions themselves. The time is past, or at least rapidly drawing to an end, when such important facilities could be sited solely by engineering and economic criteria. The new generations of major energy facilities possess a size, an importance, and a longevity that justifies the use of the best available methods for ensuring optimal locations, so that not merely feasible or convenient sites are identified. The determination of optimality is not purely an activity that should be entrusted to the utility responsible; it is no longer a narrowly defined engineering task. There has to be a tradeoff between the usual conflicting objectives of public safety, amenity, engineering considerations, and economic criteria; with perhaps an increasing degree of emphasis given to the safety aspects as a public relations measure. It is hardly satisfactory that enormously expensive investment decisions should be made in secret without any real public debate and with a minimum of opportunities for effective public participation and inputs. It is hardly satisfactory that there are no explicitly defined quantitative, or even qualitative indications of how precisely the tradeoffs were made and how the conflicting objectives were balanced. Add a belief that there is no need to engage in any detailed evaluations and comparisons of alternative sites and you have a British style deterministic siting process which is immune to either local objections, or national political control, or anything outside the narrow view of the planning engineer and the Board member.

It seems that the situation in many other countries is not much better, if at all. In the US the siting decisions are largely left in the hands of the power utilities. It is true the decision-making process is far more open than in the UK, almost painfully so because of the quasi-judicial framework within which it occurs. However, there can be little real confidence that private enterprise, driven by market forces, will be sufficiently long-sighted or benevolent to be entrusted with such fundamentally important decisions; decisions which need to be carefully managed, coordinated, and integrated if the US is to obtain the maximum benefit from the next generations of nuclear power plant.

A major focus for any rational scientific siting process must be either some form of multiattribute site evaluation methodology (Keeney, 1980) or multicriteria optimisation (Nijkamp, 1980;

11

Voogd, 1983). The methods exist and it would not require much effort to apply them to the siting problem. The problem in the UK is that there is currently no desire to use such modern techniques. In any case it would also require a fundamental change in the policy of only ever evaluating one site. It is believed that a full evaluation, in public, of all possible alternative sites would generate such public alarm and create such an intense political debate that the siting decision might never be made. However, the alternative view is that the deliberate suppression of information and the exclusive reliance on subjective assessments is also not in the best public interest, because it actively seeks to circumvent either political comment or public participation. As long as the siting process remains either a technical exercise or the commercial prerogative of a utility, and as long as civil servants and administrators have no competence to question such things, then siting will remain a largely deterministic process; a kind of rubber-stamping exercise once all the bureaucratic hurdles have been successfully negotiated and the democratic niceties observed. Instead of selecting what may be termed 'local optima' there really should be a more deliberate effort to identify 'global optimum' sites in the geographical domain of entire national spaces.

For some enterprises, this inefficiency may not matter, but for others with a potentially significant health impact it certainly does. It is true that the arguments are complex and the safety evidence neither wholly convincing nor unequivocal. Provided it is only money that is at stake then why not let the financial experts decide and argue among themselves? But when public safety is at risk, when my life is threatened and my children's future uncertain, then that is a different matter. Furthermore, it may be that far more than public safety is at stake. A major reactor accident would be accompanied by the need to evacuate probably a fairly large area which has been contaminated; perhaps for 30 years or more. If this region were to be of a low population then this sort of catastrophe would probably be acceptable; should it include high-density urban areas then the situation is unthinkable. Of course, the vested interests on both sides can both make convincing cases for their own points of view. The problem is how on earth are ordinary people, including politicians, going to judge which is the best bet. The greatest risk the pro-nuclear lobby face is this entire question of public acceptability. There are parallels here with different types of extremism. The establishment view in many European countries is that the public have no right to oppose what is considered best for them! Nuclear power, it seems, brings out the worst in those involved with it.

It seems likely then that decisions already made, and others about to be made, will determine the geography of major energy facilities for the next few centuries. The likelihood of multiple developments on the same site, of reactor parks with co-located reactors and reprocessing facilities, will ensure a high degree of permanence for selected sites. But it also means that, in practice, sites which were selected for the first generation of small thermal reactors and which are already assumed 'safe' for the current generation of far larger reactors, will almost certainly be declared suitable for the next generation of even larger reactors. Locational decisions such as these should not be made by default for reasons of convenience. There is far too much at risk should the 'wrong' sites be used.

The risks should a nuclear accident occur are twofold. Firstly, there is the threat to public health, the magnitude of which depends on several interacting variables, amongst which geographical location plays an important function. The risks appear absolutely minute but there are large degrees of uncertainty involved and the existing probability estimates may well be misleading if not wrong. The nuclear safety record looks 'good', but is it good enough? To be fair, we simply do not know. In the UK there is insufficient experience. In the US there is more experience but the safety record is somewhat damaged by the occurrence of events that should have never happened. In both countries there is not enough experience with standardised or identical plant; the duplication rate among reactors has been very small. Second, there is the risk that the financial investment in locationally fixed assets could be written off because of a collapse of public and political will for nuclear operations to continue. The following scenario might be considered plausible. Currently, the public, ministers, and administrators etc. largely believe the nuclear engineer when he says that major accidents are very rare events, and that nuclear plant is far safer than all other hazardous industries that typically cause no concern. In everyday language 'big' accidents will not occur. However, suppose such an accident does occur, perhaps in a Third World country. This implicit faith will have been destroyed. The Three Mile Island accident very nearly did this. Imagine what would have happened if 10,000 people had been adversely affected by radiation. The immediate consequence would have been pressure to close down facilities which are located in similarly populated regions. Potentially dangerous sites would probably have to be abandoned as politically unacceptable, although in practice this may not be possible, because of the capital investment that has been made

13

and the need to ensure security of supplies. In the UK nearly all the sites have higher population densities than that found around Three Mile Island.

Nuclear siting is therefore a matter of the greatest national importance the significance of which is still increasing. Site selection can no longer involve purely engineering variables, but it must become increasingly enlightened so that only sites which effectively minimise health risks and the quite independent investment decision risks are selected. The entire process must be understandable and publicly defendable since far more than a utilities future profitability is at stake. As the proportion of any nation's wealth that is tied-up in site specific developments increases, so more care and attention needs to be given to the siting process.

Is nuclear power safe?

Perhaps the most fundamental question of all is whether or not nuclear power is 'safe'. If it could be demonstrated in an unequivocal fashion that nuclear power and its related fuel cycle are completely, absolutely, and totally safe, then siting matters would be very unimportant and would concern such trivia as local landscaping, colour schemes, and the size of visitor car parks. The problem is that it cannot be so demonstrated at present. Of course we all know – partly because the nuclear industry have told us – that nothing in creation is 'completely safe'. Nevertheless, many people do regard the 'dangers' of nuclear power as being quite unique and, rightly or wrongly, as being at least qualitatively different from perceptions of all other dangers that are normally associated with human existence on earth in the 1980s. Many of these fears may seem irrational and unscientific but they are often very real and cover various combinations of big accidents, nuclear sabotage, terrorism, nuclear weapon proliferation, state tyranny and plutonium economies, unidentified risks associated with low level radiation exposures, and cumulative environmental pollution with radioactive material. It may also reflect a feeling of helplessness that major, new, perhaps risky or 'dangerous', nuclear power developments can be seemingly pushed through regardless of local opposition. In the UK it has proven virtually impossible to argue successfully the local case when national interests are at stake; this applies to nuclear power stations and gravel pits. Nuclear electricity is a very strong vested national interest. Without it, it seems civilisation will almost instantly be plunged into a sort of stone-age subsistence economy! Nevertheless, it is only going to be acceptable if it is 'safe' and siting will only really become important if it is not.

The nuclear industry now sees the problem as an urgent matter of public education (some people might say conditioning) so that the 'risks' involved can be properly appreciated (dismissed as trivial) by viewing them in the most appropriate (most favourable to the nuclear industry) context. One cannot help having considerable sympathy for the dilemma of a nuclear industry which, when faced with increasing public awareness of the potential problems of nuclear power, has to try and explain that whilst their safety cannot be guaranteed and no absolute assurances can be given, nonetheless all fears, suggestions, and perceptions that nuclear power is unsafe are ridiculous and not based on the facts. This may not be an entirely new public relations problem but it is an emerging issue.

Historically, public apathy and security restrictions largely precluded public debate, and the number of dissenting people were limited to a small lunatic fringe. The public were either not interested, or were easily reassured by expert statements, or regarded nuclear plant with civic pride as a status symbol of the jet age. Today the situation is quite different. There is a broader appreciation of the seemingly growing problems of environmental contamination following various harmless and health insignificant leaks of radiation. In the UK, the problems of Seascale have been given massive press coverage. So what if 15 miles of beach were contaminated by a small leak of high level waste, or that the estuarine silts in the neighbouring harbours contain so much plutonium that samples taken for analysis cannot be disposed of at civilian dumps? The point is that the observed health effects are seemingly and at present virtually nonexistent. However, the attention given to Three Mile Island, the possible Russian disaster of the 1950s, and the Sizewell B PWR Planning Inquiry of 1983-5, have made it necessary for the public and the politicians to be reassured that all is still well.

The problem today is no different from the early days of nuclear power. Lord Hinton of Bankside, who was responsible for the development of the first generation of UK reactors and was later the first chairman of the CEGB, neatly summarised the basic safety problem when he recollected a statement he made in Dounreay town hall early in 1955 (Dounreay was the site selected for a prototype fast reactor that some people feared might go promptly critical and explode). He said:

> Let me make it clear that I am not going to claim that there
> are no risks associated with the development of Atomic

Energy. Every human activity involves a certain measure of risk and we can only evaluate one risk by comparing it with other additional risks to which we are subjected (Hinton, 1977, p. 12).

He quantified the risk of death from natural causes for people aged between 10 and 45 as 150 per 100,000 people per year. Of this 150, nearly 40 would be due to accidents, and less than 3 would be due to the additional risks of working in an atomic energy establishment. Since this time, the same argument has been greatly extended to put the additional risks of nuclear power in its proper context. However, there are two major problems:

1 the resulting death risk probabilities forecast by the nuclear industry are so small that lay people have great difficulty in understanding what they mean;
2 few people outside the nuclear industry seem to believe them, even the experts appear to be divided as to whether such estimates are either realistic, meaningful, or accurate and a number of former nuclear engineers have expressed grave doubts.

In the UK the CEGB have design guidelines for their latest reactors which suggest that severe reactor accidents resulting in large uncontrolled releases of radiation should have a frequency of less than once every 10^7 years (Mathews, 1982); which criteria are applicable for the earlier stations have never been declared. In the US it seems that similar accidents are expected once every 10^4 years of operation for reactors which are essentially the same as planned for the UK – British readers will immediately recognise that this improvement in safety is a result of skills of British designers, craftsmen, and reactor operators! In fact both sets of figures might be regarded as pessimistic when they are converted into annualised death risk probabilities. In the UK, a person living within 10km of a reactor faces an additional death risk of 1×10^{-11} per year; the problem is that he or she probably does not know what this means.

The 10^{-7} probability indicates that severe uncontrolled releases of radiation with the possibility of high casualty rates may be expected to occur once every 10 million years of reactor operation. It does not mean, however, that with 10 reactors in operation it will on average be 1 million years (1001985 AD) before an accident occurs. Such an accident could happen next year or even today although no one really expects that such low probability events will actually happen. Any sane person would probably be justified in equating this magnitude of risk with zero. However, a problem with numerical safety criteria is that once they have been computed it is difficult to drop the spurious levels of precision with which they are specified.

It is fairly straightforward to place these accident probabilities into context by comparing them with various other risks that people face merely by being alive in the 1980s. Table 1.1 gives a representative comparison for England and Wales in 1978. In that year it seems that none was knowingly killed by radiation accidents.

TABLE 1.1 *Annual death risk probabilities for England and Wales in 1978*

Cause of death	Probability of death per person per year
All accidents	.00044
Motor vehicles	.00014
Falls	.000095
Suicides	.000082
Accidental poisoning	.000016
Fire and Flame	.000013
Murder	.000012
Accidental drowning	.0000072
Falling objects	.0000022
Rail travel	.0000025
Electric shock	.0000026
Water transport	.0000024
Air travel	.0000010
Firearms and explosions	.00000086
Lightning	.00000002
Earthquakes, floods	zero
Reactor accidents	zero

Source: OPCS (1980).

Beckmann (1976) provides an even more convincing demonstration. He argues that a better way of gauging the danger is to look at the mean number of fatalities per year and compare them with other accidents. To prevent nuclear power from looking too good he takes average probabilities from WASH 1400 (NRC, 1975) applied not to the whole US population but to the 15 million Americans who live within a 25-mile radius of current and planned reactor sites. The results are shown in Table 1.2.

TABLE 1.2 *Expected annual fatalities among 15 million people living within 25 miles of US reactor*

Cause of death	Number
Motor vehicles	4,200
Falls	1,500
Fire	560
Electric shock	90
Lightning	8
Reactor accidents	2

Source: Beckmann (1976), p. 70.

Further reassurance can be offered by noting that even worst-case reactor accidents seem to be similar to other possible catastrophes and cataclysmic events that we all have to live with. The WASH 1400 report compares the risks of nuclear accidents to other remote events such as meteorite impacts on cities, earthquakes, aircraft crashes, dam failures, and fire. Typical frequency-casualty relationships are shown in Figure 1.1. According to the WASH 1400 report the risks of a 100 reactor power programme are far smaller than all other man-made events with similar numbers of casualties (see curve A). Indeed, it would seem that many cataclysmic events with a similar casualty potential as reactor accidents are in fact 1000 times, or more, more likely to occur.

Whilst seemingly exceptionally favourable to nuclear power, it seems that displays such as Figure 1.1 are self-defeating, because it reinforces the idea that the large numbers quoted for hypothetical reactor accidents are all sudden deaths just as they are for earthquakes or dam failures (Marshall *et al.*, 1983). Another criticism might be that it gives the impression that other forms of large scale energy conversion are safer than nuclear power or that large accidents are uniquely associated with nuclear power. There are those who argue that if a big nuclear accident is even remotely possible then we should not have nuclear power stations. The standard rebuff is to take a parallel point of view in a different field – for example, civil aviation. The question can be asked, 'In view of the possibility of serious flying accidents, should civil aviation be allowed?' (Marshall *et al.*, 1983, p. 8). It could be pointed out that a Jumbo Jet might occasionally be expected to crash and kill 300 people; it could crash with full tanks on London's Wembley Stadium when it was full of people

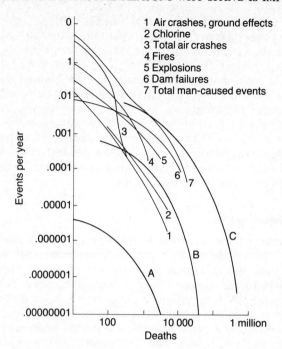

FIGURE 1.1 Accident frequency and casualty size relationships for man-made events
Source: compiled from various figures in UCS (1977) and NRC (1975).

and kill 10,000 or even 50,000. It is not impossible but it is very unlikely. It has a probability of about 10^{-8} per year, but no one has ever argued that planes should not be allowed to fly over football stadiums.

This line of reasoning can be extended further to include the relative risks of nuclear and conventional power generation. As Beckmann (1976) points out: 'Not only are major accidents with fossil fuels and hydropower far more probable, but their consequences can be more terrible too' (p. 63). There is no need to imagine Jumbo Jets crashing on to full football stadiums since explosions at oil and gas storage facilities have already killed hundreds of people and it is not difficult to imagine even higher death tolls.

However, before becoming too convinced about the safety of nuclear power stations it is important to realise that Figure 1.1 is now generally regarded as being wrong. The Union of Concerned Scientists have attempted to re-assess the risks predicted in the

19

WASH 1400 report by correcting various methodological peculiarities. They discovered that when the obvious errors and omissions had been rectified, the consequences could be 100 to 1000 times greater (UCS, 1977, p. 113). Curve B in Figure 1.1 shows the revised frequency-casualty line for prompt deaths. If latent deaths are added then curve C might well be considered as being more representative of the nuclear power risks for comparison with earthquakes. Indeed the situation is now very different in that over the complete range of large casualties (more than 1000) nuclear power presents the greatest risks. Furthermore, the uncertainties in these estimates are large, so that even curve C might not be an upper limit.

The radiation game

The public relations problem concerns the need to explain precisely how safe nuclear power is in terms that are both realistic and readily understood by the public at large. One solution is to use simpler ways of explaining what the probabilities mean. A very interesting report has been produced by Marshall et al. (1983) which attempts to correct what the authors consider to be the general public's distorted impression of the risks of nuclear power due, apparently, to a widespread, irrational fear of hypothetical, very unlikely, gigantic nuclear accidents; or perhaps even what may be termed media fantasies. This study is very important because the senior author is Sir Walter Marshall, one-time chairman of the UK Atomic Energy Authority, author of the Marshall Report on the safety of the PWR (Marshall, 1976), and currently chairman of the Central Electricity Generating Board. Their recommended strategy for laundering the nuclear power industries tarnished image is as follows. First exclude from graphs (such as Figure 1.1) and casualty estimates the prompt deaths, victims (if there are any) who would be unequivocally attributable to a reactor accident – although it might be argued that some deaths would in any case be due to natural causes. The numbers involved here would be very much lower than the potential number of latent or long-term cancer deaths, which is viewed as being the basis for the misleadingly large numbers of casualties often quoted in newspapers. For example, the delayed cancer deaths predicted in the German Risk Study (1981) number 104,000 for an imaginable reactor accident, or the 36,000 cancers predicted by Hemming et al. (1983) for a large radiation release at Sizewell under extreme weather conditions. These latent death, or long-term fatal cancers can be more accurately described as '. . . adverse health effects leading to premature deaths from radiation induced cancers over a period of 30 years starting some 10 years after the accident' (Marshall *et al.*, 1983, p. 12). It is considered misleading to quote potential long-term deaths and the short-term deaths as if they are the same. Whether the public would take such a tolerant view is doubtful, especially as they may well be cancer-adverse in their risk-taking because of their fears of that illness, and because most cancers are not uniquely attributable to radiation so their chances of receiving any compensation would be small.

Marshall *et al.* (1983) recommend the use of three related measures which are designed to overcome the problems of the public's misunderstanding of what total cancer predictions really mean. By making a few simplifying assumptions they derive general formulae for computing the loss of life expectancy to a population exposed to radiation, a means of estimating their increased probability of dying from cancer, and a way of expressing the adverse health effects they may suffer in terms of an equivalent pattern of compulsory cigarette smoking. This study is sufficiently important, and the ideas it contains are sufficiently characteristic of other attempts to explain how safe nuclear power really is – as viewed by those engaged in it – that a fairly lengthy discussion is useful.

As far as the loss of life expectancy is concerned, Marshall *et al.* (1983) conclude that a dose of 1 rem is equivalent to a loss of life expectancy of 20 hours. They somehow seem to believe, or assume, that 1 rem dose is all that might be expected following a gigantic, unlikely, or highly imaginative 'big' reactor accident. They write:

> It is therefore fair and proper to explain that the consequence of a hypothetical but imaginable accident is to reduce the life expectancy of a person who could have been affected, on AVERAGE, from something like 38 years to something like 38 years less a few hours or, for an extreme accident, less a few days. It is therefore fair to explain to the public that we do not expect big reactor accidents to happen, the chance of their happening is extremely remote and that, *even if one did happen*, then ON AVERAGE their life expectancy would be reduced by this small amount due to the adverse health effects of the radiation released during the accident (p. 47).

It should be noted that a UK citizen of average age would at any moment in time have an average life expectancy of 38 years. This sort of statement is exceptionally misleading. Even Marshall *et al.* then explain that these average results do not properly describe the stochastic nature of the risks, having previously stated that the statement about loss of life expectancy is both quite precise and correct. They declare that:

> . . . when using loss of life expectancy . . . we must be careful not to mislead members of the public into thinking that they might lose a few hours of their life. We must explain that most people would be unaffected but some, the unlucky ones, might lose years or decades of their life (p. 49).

This qualification contradicts the previous unequivocal statement and still stops short of identifying the very unlucky people who lose all their remaining life expectancy. So clearly this way of presenting the worst dangers of nuclear power is unsatisfactory.

The second approach is to measure the effects of a reactor accident in terms of an increase in an individual's cancer probability. The following assessment is made:

> For each rem an individual receives from a hypothetical accident, this cancer probability is increased by $1/30$th of 1 per cent (i.e. an absolute increase in cancer probability of $1/150$th of 1 per cent). In other words, even if an imaginable accident were to happen, the adverse health effects are an imperceptible increase in an individual's normal chance of dying of cancer (p. 49).

Again this is only true if a 1 rem dose is all that is received and it still requires a person to have a feeling for the smallness of $1/30$th of 1 per cent or of the magic ICRP dose response coefficient of 3.4×10^{-4}.

The third and final suggestion is to re-examine the problem in terms of compulsory cigarette smoking. For some reason, Marshall *et al.* consider that it is sensible to compare the risks from nuclear power with the risks from cigarette smoking, presumably because they perceive the magnitude of the risks to be similar and both cause cancers. The distinction between voluntary and involuntary risks is ignored. They suggest that:

> . . . the risk from a radiation dose given to an individual at any age is best simulated by supposing that the individual is obliged to smoke cigarettes regularly starting 10 years after the incident and ending 40 years after the incident. . . . We are now able to explain to the population that the risk from a radiation dose of 1 rem is the same as the risk from being obliged to smoke approximately $1/20$th of a cigarette every Sunday. This level of cigarette smoking is so low, it must be less than the involuntary level of cigarette smoking forced on

them because other people smoke alongside them . . . (pp. 50-1).

The analogy is regarded as most useful because the public supposedly have a good appreciation of the risks of cigarette smoking, and because it does not require a precise numerical expression. They might just be right but it is exceptionally misleading to continually convey the false impression that the health consequences of a gigantic reactor accident, presumably the worst 'big' accident that the chairman of the CEGB can imagine, can be realistically assessed in terms of a 1 rem population dose. They go on to consider the effects of such a 'big' accident on the 10 million people in the London area that again gives everyone a 1 rem dose! According to Marshall *et al.* there would be no short-term deaths, and although the long-term death toll is 1,250 people all but 570 people would have died from other causes anyway. Instead of using these numbers which might look bad, they would explain that the effects of such an accident would be to reduce the average Londoner's life expectancy by less than one day, their chances of dying from cancer would be imperceptibly different from what they were anyway, and that the adverse health effects are equivalent to everyone in London compulsory smoking about $1/20$th of a cigarette every Sunday.

A recent report by the independent National Radiological Protection Board has, as it happens, discussed the impact of a 'big' reactor accident at Sizewell on the London area (Kelly *et al.*, 1983). They assumed that the whole area would be evacuated within 2 days of the accident happening. Even then, there is a wide range of possible early and late adverse health effects depending on weather conditions. Early deaths from a typical hypothetical accident varied from 0 to 860,000, and late effects from 56 to 440,000, assuming no evacuation – which is probably the most realistic assumption. If 10 million people could be evacuated within 2 days then there might only be between 1-150,000 deaths and 420 to 40,000 fatal cancers. It should be noted that London is at least 120km from the Sizewell reactor. Who should we believe?

Whilst one must sympathise with the problems of explaining to the public matters of nuclear safety, it is not possible to condone the use of potentially misleading analogies, especially when there is no satisfactory empirical basis for the risk equivalences between compulsory cigarette smoking and radiation. The tobacco industry still consider the link between cancer and cigarette smoking to be not causally proven, and there are no grounds for believing that the dose response relationship between the number of cigarettes consumed per week and cancer is linear. A radiation dose of 900 rems might be expected to kill a person fairly quickly if it is received over a short period of time. The equivalent compulsory cigarette smoking would be 45 per week starting 10 years after the accident and then continuing for 30 years at which time, if the analogy were a good one, the smoker would be dead. One imagines that this analogy only works for small radiation doses which are probably not relevant for considering the effects of 'big' accidents; even then, non smokers might find it offensive.

In a 'big' accident the number of people exposed to some excess radiation, even if it were a very small dose and totally insignificant, would be very large. It is inevitable, therefore, that the *average* effects turn out to be so small as to give a false sense of security. Consider cigarette smoking. The individual risk per cigarette is the number of excess deaths attributable to smoking divided by the number of cigarettes consumed. For the US these figures may be of the order of 300,000 excess deaths per year divided by 591 billion cigarettes, giving an average risk of 4×10^{-7} per cigarette consumed. This very small individual risk factor completely hides the impact of the real magnitude of the excess mortality caused by cigarette smoking. Does this very small number mean that there is a negligible personal risk? No – indeed various national governments print a health warning on each packet! Clearly, then the compulsory cigarette smoking analogy is equivalent to HM Government including a health warning on each electricity bill; but unlike cigarette smoking, even if you stop or reduce your electricity consumption the risk still remains.

So it seems that the misleading impression the general public may have about the adverse health effects of BIG reactor accidents can be explained away as being due to their misunderstanding of numbers. What is perhaps more worrying for the public at large is that the stochastic component in delayed health effects can be used to assuage blame or liability for the majority, perhaps all, long-term cancers. The unlucky victims of long-term health effects, or should I say, possible victims, would have to prove that their injuries were uniquely due to a reactor accident perhaps 10 perhaps 30 years previously. The 'official' estimates of radiation dose may well be so low as to indicate that no blame could be attached to the reactor accident; alternatively, there may be no accurate estimates of dose. Even if the affected persons kept lifetime dosimetric records the evidence would still be circumstantial. In short, it may be thought that the public should not rely too heavily on the explicit assurances of safety that the nuclear industry publish; they would not dare say anything else. If a 'big' accident ever occurs, and the risks may well be small, the last thing the nuclear industry concerned would do is admit unlimited liability. Such a policy is only common sense; but it could also amount to a deliberate exploitation of the stochastic component in radiation-induced illnesses, or of the lack of concrete data on actual doses as received by the public.

At present the public must rely on the safety measures taken by the nuclear industry to protect them from the involuntary risks that they impose on the public in the best interests (according to their point of view) of the community. However, it would appear that there is little hope of compensation should things go wrong, unless someone admits liability. But on the other hand, the nuclear industry should not expect the public to judge nuclear power to be acceptable merely on the grounds that the average impact of the technology is comparable to, or even smaller than, the average impact of competing technologies (MITRE, 1977). Indeed, it is far more realistic to assume that the public assessment of nuclear power is that it is markedly different from other harmful technologies, and that accordingly it must also be markedly safer, irrespective of the actual levels of risk that can be measured.

But once again the evidence is contradictory. The overall societal implications of nuclear power compared with the alternatives are far less. Beckmann (1976) estimates that the number of excess deaths due to respiratory diseases are between 20 and 100 per 1000MW of coal-fired plant per year, or that every 1000MW of nuclear power can save this number of lives per year. Chicken (1981) estimates that the cost in human lives of using coal-fired power stations instead of gas or nuclear power will be tens of millions of pounds per year; or at least it would be if the victims could prove they were victims. Table 1.3 gives estimates of fatalities per year for different types of power generation. The problem here is that there is no accurate way of making these

TABLE 1.3 *Estimated fatalities per 1000MW(e) power station*

fuel	Pochin (1976)	Hamilton Manne (1977)	Inhaber (1978)
coal	45	10-200	50-1600
oil	1.5	3-150	20-1400
gas	0.1	0.2	1-4
nuclear	1.0	1-3	2-15

sorts of statements for conventional plant, whilst the nuclear casualties reflect the dangers of uranium production but not the risks of either reprocessing or big accidents. However, even if the effects of big accidents were to be included then it may not be noticeable, since whatever casualty rate was used would be assumed to be a once every billion years event.

There is also an interesting philosophical question that needs to be answered. Comparisons of risks and of alternative technologies are philosophically weak because they imply that one activity is better than another, merely because the number of fatalities associated with it represents either a small deviation from the current death rate or because it reduces the death rate by a small amount (Shrader-Frechette, 1980). The question is basically what is considered to be 'normal' for comparison purposes, and what the risks should be without considering what they actually are. This problem is particularly dangerous within technology assessment because '. . . it allows one to beg important questions and merely define them in terms of past answers. . . . As a consequence, whatever social and ethical discrepancies that are present in the status quo, are also present in the technology assessment. . . . Hence the views, of whatever people set the norms, enjoy an undeserved and therefore irrational power' (Shrader-Frechette, 1980; p. 151). No additional comments are necessary.

A final aspect of the radiation game is known as the zero-infinity dilemma. If the probability of an accident is exceedingly small but the consequences are exceedingly large then the task of making a meaningful risk assessment becomes impossible. Is the risk negligible because the event may never happen, or is it large because the consequences are so great? There is no consensus as to how to solve this dilemma. It is in this context that the size of the worst disaster that can reasonably occur becomes so important. Precisely what the consequences are of 'the worst accident' that could happen have never been properly defined. Most estimates of casualties have been based on remote accidents defined by reactor manufacturers. The point to note here is that an imaginable worst-case accident could result in: the large scale evacuation of a very large area for a long time; restrictions on the consumption of food and water over an even larger area for a long time; a casualty rate that is a direct reflection of the population density of the affected regions and the time taken for evacuation; and a cancer rate that may be impossible to accurately measure because it could have a global component to it. For the Sizewell reactor the area to be evacuated might have to be the whole of Greater London and other parts of the South-

East. An accident at a coal-fired station could never cause anything like this scale of disaster. Hence the zero-infinity paradox of nuclear power.

The probability of such a large accident may well be vanishingly small, but the consequences could be so devastating. Additionally, it should be noted that similar low probability events do occur and, moreover, might well be expected to occur in the immediate future. For example, if the probability of a major reactor accident is 10^{-6} per reactor year: if you have 50 reactors in operation for 40 years, then a little arithmetic suggests that on average 1 major accident might be expected every 500 years. If worldwide there are 1000 reactors in operation, then the average time between accidents reduces to 1 every 25 years. In this situation it can be argued that the fate of one country's nuclear power programme is not independent of the rest of the world, so a serious accident anywhere could well have repercussions far beyond the country where it occurred.

Until nuclear reactors with a zero accident probability can be built then it seems that nuclear power is a distinct gamble that is probably not in the best interests of either consumer or power utility. At risk is not merely one's life and property but also the future levels of electricity generation, and hence probably also the future of an advanced nuclear powered civilisation. The whole business is a gamble with unknown odds, and with no real appreciation of what is at stake should the gamble fail. A prudent approach would, even with existing a priori risk estimates, seek to plan for the worst accident by using the geography of reactors and people as a combined safety and public relations measure. If accidents do occur then the public and the politicians may well be surprised that so little attention has been given to geographical isolation as an independent and complementary safety measure.

Why worry, the predicted accident probabilities are so vanishingly small

It is true that the risks in Table 1.1 for reactor accidents were zero in 1978. Even if you assess the effects of degraded core accidents at Sizewell the average annual personal death risk increases to between 3.1×10^{-7} to 3.3×10^{-11} (Kelly and Clarke, 1982). These numbers are so small that they are just about finite approximations to zero. They are small largely because severe accidents are assumed a priori to have a probability of occurring that is also very small; the 12 accident scenarios evaluated at Sizewell had frequencies of occurrence from 5.1×10^{-7} to 2.0×10^{-10} according to the reactor designers (Kelly and Clarke, 1982).

The question is, where do these estimates of reactor accident frequencies come from?

A guide to the origins can be found in developments in the UK in the 1960s. Farmer (1967) suggested a hypothetical frequency casualty relationship as providing a useful basis for a probabilistic safety criteria. He argued that no engineering plant is entirely risk-free and that there is no logical way of differentiating between 'credible' and 'incredible' accidents. The only satis- factory approach is to examine the whole spectrum of risks in a quantity related manner. A measure of risk can be obtained by estimating the probability of a failure and then assessing the consequences. A full safety evaluation would comprise a spectrum of events with associated probabilities. A probability consequence diagram could then be drawn. All parallel lines with a slope of -1 identify points of equal risk; see Figure 1.2. Farmer (1967) speculated that one such line might be identified as an upper boundary of permissible probabilities. Reactors could then be designed to conform to such a curve.

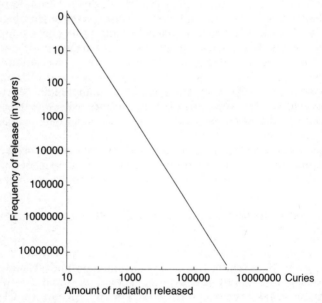

FIGURE 1.2 Probability consequence diagram in the style of Farmer (1967)

To allow for the fact that lines of equal risk may not represent lines of equal risk of casualties, Farmer suggested a line with a slope of -1.5 so as to reduce by three orders of magnitude the frequency of an event whose severity increases by two orders of

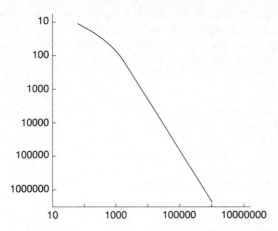

FIGURE 1.3 *Modified probability consequence diagram to reduce frequency of small releases to avoid public anxiety*

magnitude; see Figure 1.3. The position of the line reflects: possible public reaction to an accident, the estimated numbers of casualties, and the increased risk incurred by any individual. The top part of the line (for releases less than 10^3 curies) is curved to ensure that the risk of one event by 2000 AD when 1,000 or so reactor years of operation will have accumulated in the UK is less than 1 in 3. The curve between 10^1 and 10^3 curies is meant to minimise the frequency of small releases which might be considered to have a nuisance value.

The idea is that reactors can be designed to satisfy a Farmer-type target line. Fault tree and event tree analysis would be used to identify events or combinations of events that might result in accident frequencies greater than the target line and additional engineering safeguards would be introduced to provide the desired degree of safety and reliability. This probabilistic approach was strongly recommended by Flowers (1976) but it was not clear as to whether it has actually been used in practice, and doubts have been expressed about the utility of fault tree and event tree analysis as means of identifying unexpected combinations of events. It would work best if all accident generating sequences could be identified, but this is of course impossible to achieve with complex systems; there will always be something important missing or there will be some assumptions which may not hold good under all practicable conditions.

The Farmer target line concept is based on the principle 'the bigger the accident the less likely or less frequently it should happen'. It also implies that there is no upper limit on consequences. Finally, and this is important, it reflects a value judgement about the sort of risks that it would be reasonable to expect society to tolerate. Cottrell (1981) expresses the problem like this:

> . . . if civil nuclear power killed people in the same numbers and frequencies as people are killed by aircraft falling on them, would this be publicly acceptable? At this point we have to make a judgement. Even compared with aircraft, nuclear power is a new hazard and so cannot expect to enjoy that degree of acceptability which aircraft have now gained out of sheer familiarity, through people having 'learned to live with them' (p. 73).

So some higher safety factor is required but what would be acceptable? In a world of 1000 reactors would a risk $^1/_{10}$th that of aircraft be acceptable? It may imply a catastrophic reactor accident once in every 10,000 years with about 1000 casualties and a single death every 10 years. Would this be acceptable? Cottrell (1981) suggests:

> We can only guess that it might, since it specifies a level of safety distinctly better than that of the least hazardous of the other man-made accidents . . . and because it provides on average a large gap in time between casualties (p. 73).

But what is the implication for individual risk rates? The target line concept can be converted into individual probabilities in the following way. If 10^7 curies are released once every 10^6 years with say 10,000 deaths, then it is assumed that this is equivalent to killing one person every 100 years or 5.2×10^{-9} of a person each year.

There are at least three flaws in this sort of argument. First, how do we know that a live reactor actually conforms to a specified target line; or indeed, what target line the reactor is supposed to conform to? Second, the assumed reactor accident risk factor may be grossly underestimated. Farmer (1967) considered that a 1 in 3 chance of a 1000 curie accident by 2000 AD was tolerable in the UK. On the other hand, ten years later the Union of Concerned Scientists estimated that if the hazards presented by a programme of 100 nuclear stations is constant through time (a big if), then there is a 1 per cent chance of nearly 100,000 fatalities and a 0.1 per cent chance of 180,000 fatalities by 2000 AD (UCS, 1977, p. 144). Thirdly, it is misleading to quote individual risk rates for reactor accidents which are, by their very nature, area events. A car accident is normally something that affects a small number of people. It is not useful to compare car accident risks with nuclear power risks which affect not a few people, but in their extreme form which is all that is of interest, many people more or less simultaneously. If the probabilities are corrected to take this into account then the interpretation from Table 1.1 would be very different.

Consider an example. The risk of every person in Heysham, a town located very near to four AGR reactors, being killed in any one year in car accidents is about 2.7×10^{-168000}. However, the probability of them all being killed in a reactor accident is virtually the same as one person being killed, since an accident severe enough to kill one person will probably be sufficiently severe to wipe out the entire town. The risk of one person being killed is approximately 2.5×10^{-9}, so to the inhabitants of Heysham, the risks of death due to nuclear power is about 2×10^{167991} times greater than the risks presented by traffic accidents! This could be compared with the impression that the car is about 10^5 more dangerous than nuclear power in Table 1.1. So it seems that probability arguments can be readily manipulated to prove virtually anything.

Other ways of obtaining 'small' nuclear power death risk probabilities are a combination of the chain multiplication of probabilities and averaging. Take a typical reactor accident scenario with an estimated frequency of occurrence of 9.5×10^{-11} per reactor year, according to the manufacturers. The health consequences, and thus the individual deaths risks, depend on such variables as wind direction, wind speed, atmospheric stability, and rain out. So establish for a site (perhaps using meteorological data from 50 miles away) the quantity $C(i,m)$, where C is the health consequence for wind direction i and meteorological conditions m. The probability of i and m occurring is $f(i,m)$ and the probability of an accident with a particular radiological profile occurring in any year is $F(r)$. The average individual risk is the product of these quantities divided by the size of the affected population, viz.:

$$\frac{c(i,m) \times f(i,m) \times F(r)}{\text{population}}$$

It can be disaggregated by distance bands and sectors and will generally produce minute accident probabilities.

At face value this seems to be a reasonable approach and procedures of this type are currently the basis for casualty estimates in both the UK (Clarke and Kelly, 1981) and the USA (Ritchie *et al.*, 1982). There are a few methodological problems

31

concerning the representativeness of the samples of weather sequences that are used and the nature of the F(r) assumptions which are taken at face value. It may indeed be reasonable to average consequences over all weather types even if such results are biased against more infrequent weather conditions. So it may be necessary to assume that those rare accidents occur at times of average weather conditions. The possibility of rain out is particularly important and has the risk of major health consequences, but the distribution of rain is something that current computer models largely ignore. This is understandable because modelling it requires detailed three-dimensional weather data on a time lapse basis and such data are generally not available on a sufficiently fine geographical scale to be of any assistance. Other problems relate to the use of night-time population data and the use of atmospheric models with uncertain parameter values. Until a major accident occurs the models cannot be validated except in a most approximate fashion.

Where DO the reactor accident frequency estimates come from?

So far it has been accepted that the frequency of large nuclear accidents is vanishingly small; but how do we know that they are so small and with what confidence can such estimates be made? The answer might be expected to be contained in various reactor safety studies, except that in the UK none have been published (yet). Very little is known about the levels of safety attained by the MAGNOX and AGR reactors, other than that they are incredibly safe (Macdonald et al., 1977) and it is only recently that any details of their design have been published (Dale, 1982). For further information it is necessary to look at various US reactor safety studies.

The first of these, known as WASH 740 (USAEC, 1957), is probably the first attempt to assess the effects of a reactor accident that has been published. The results are slightly horrifying. It was estimated that one of the worst possible accidents at a 200MW nuclear reactor situated 30 miles from a major city could result in 3400 deaths, 43000 injuries, and 7 billion dollars' worth of property damage. No estimate of the probability of such an accident could be made. The principal result was that no utility was thought likely to be interested in nuclear power unless the US government would indemnify them; indeed the subsequent Price Anderson Act limited their liability to 560 million dollars, of which government underwrote 500 million (Brown, 1976). In the UK, the CEGB's liability was set at the lower limit of £5 million sterling. The Price Anderson Act was repealed in 1982, whereas that in the UK is still in force. So in the event of a nuclear mishap, individual householders have no recourse to their own insurance firms and the utilities responsible had or still have a limited liability; the excess may or may not be paid by government.

In 1964 the US Atomic Energy Commission initiated a new study to update the WASH 740 results by taking into account the larger plants now being considered and the introduction of new analytical techniques. It was hoped that the analysis would now demonstrate that the WASH 740 estimates were pessimistic. The revised results suggested that an accident at a 1000MW plant located in a city could involve 45,000 fatalities, 70,000 injuries, and a possible disaster area the size of the state of Pennsylvania. The results of this 1964-5 study were never published until a Freedom of Information request in 1973.

The next major safety study was the WASH 1400 report (NRC, 1975). This was a detailed attempt to determine both the probability of nuclear plant accidents and their likely consequence in terms of both personal injuries and property damage. The study was only concerned with the risks from power reactors (it excluded the fuel cycle) and was considered applicable only to the first 100 light water reactors and to operations in the period 1976-80. It made use of techniques developed for the aerospace programme to estimate accident probabilities. The consequences of a major accident to a 1000MW reactor ranged on average from 3,300 to 45,000 deaths; 45,000 to 248,000 injuries, and property damages from 14 billion dollars upwards to an unspecified upper amount. The general conclusion was that a major accident resulting in the dispersal of large amounts of radioactivity is a 1 chance in every billion years of reactor operation.

The WASH 1400 report uses fault tree and event tree analysis. It offers a systematic approach to identifying hazardous accident sequences, assigning them numerical probabilities, and comparing the likelihood of different sequences. Methods like this have been widely used in the US aerospace programme. However, whilst it was useful as a general means of assessing the impact of design changes and as a means of identifying particular failure consequences, it was subsequently abandoned as a means of measuring the safety of a system and as evidence that a complete safety analysis had been performed. It was not considered feasible that all important accident modes could be identified and its use as a basis for predicting accident probabilities is now considered technically indefensible (APS, 1975). At best, attention has to be restricted to credible events and the recognition of

33

what is plausible is subjective. The Union of Concerned Scientists (1977) explain:

> . . . one must have confidence that all important accident sequences have been identified. The fact that one has come close to disaster in a couple of hundred years of experience, and had luck in not going any further, belies claims of likelihoods of one in a million or less per reactor year (p. 18).

In fact, the UCS (1977) then demonstrated that at least one near reactor accident in the US (at Dresden II) occurred, but according to WASH 1400 had a predicted probability of $2.4 \times 10^{-3.8}$ (p. 186). Clearly it should never have occurred, but it did.

Other criticisms of the WASH 1400 report include: a neglect of common mode failures in which the accident sequence itself initiates other failures; doubts about whether pressure vessel failure deemed incredible can in fact occur; the neglect of the effect of age and weakened components on failure rates; the exclusion of sabotage, intentional accidents, and operator errors; and doubts about the effectiveness of emergency plans when the public are not included in mock drills. The UCS also estimate that the consequences could be 100 to 1,000 times greater. In particular it was concluded that nuclear risks appear to be substantially greater than many of the risks that are man-caused:

> For events with fatalities above about 2000, nuclear plant accidents can exceed the sum of all other man-caused risks, and the largest nuclear accident considered could have a very appreciable impact (p. 114).

It now seems that the chances of a major light water reactor accident somewhere in the world before 1990 are estimated at between 1 in 3 and 1 in 5 (HCSC, 1981).

It can be argued that until assurances are given that no accidents will happen there should be no large-scale commitment to nuclear power. Even a small accident, with a few hundred fatalities, could cause public pressure to curtail the operation of nuclear plant and seriously disrupt electricity supplies. Even the occurrence of a single accident exposing 300,000 or so people to differing levels of radiation with only 1,000 deaths – much less than the average WASH 1400 accident – would probably be a fundamental obstacle to a continuing nuclear power programme. Every person living near to a reactor would probably think they might be next. The risks are at a level in the US where a modest accident has an appreciable chance of occurring by 2000 AD. An expanded UK nuclear power programme would bring with it a similar level of risks. The gamble according to the UCS is that:

While we cannot exclude the possibility that the country *could* escape such an event in the foreseeable future – because the uncertainties are so substantial – it is far from plausible that the country *will* (UCS, 1977, p. 130).

The problem in the UK is that US experiences are not regarded as being relevant. The safety points raised in the Kemeny Report following the Three Mile Island incident are regarded as irrelevant. The UK Health and Safety Executive reported that 'The comments in the report are not relevant to the situation in Britain and Europe more generally . . .' (HSE, 1981, p. 21). The reasons were not given.

The CEGB have an excellent safety record

In the strange world of the nuclear industry's public relation programmes, Three Mile Island is now regarded as a major plus for the safety of nuclear power. No one was killed despite a major accident. This absence of directly attributable deaths is often used to indicate how safe nuclear power really is when compared with conventional plant. A typical UK statement is that by Matthews (1981):

> The CEGB's operational experience of nuclear power dates back to 1962 and now amounts to over 200 reactor years. During that time there has been no evidence of any harm being caused by radiation to any worker or member of the public (p. 16).

The implication is that this is supposed to prove that *all* nuclear power plants are safe. Such an interpretation is not justified simply because the data base is too small and the experience referred to by Matthews is based on a number of different, usually prototype, reactors. At best, all it proves is that the likelihood of a cataclysmic accident is probably not more than 5×10^{-3} per reactor year. This it could be argued is far below the level of safety needed to show that nuclear power is safe.

Well within ICRP recommended limits

An important aspect of the general safety situation is the reliance that is placed on conformity with the International Commission on Radiological Protection (ICRP) recommended radiation standards. Maximum permissible doses are set for radiation workers and the public. All radiation from nuclear sources can be harmful and the ICRP limits represent a balance based on an assessment of risks and benefits. In their 1966 Recommendations they write:

> This limitation necessarily involves a compromise between deleterious effects and social benefits. . . . It is felt that this level provides reasonable latitude for the expansion of atomic energy programmes in the foreseeable future . . . the limits may not in fact represent the proper balance between harm and probable benefit (ICRP No. 2, 1959).

They recommend that all exposures be reduced to either the 'lowest possible level' (1955); or to 'as low as practicable' (ICRP No. 2); or to 'as low as readily achievable, economic and social considerations being taken into account' (ICRP No. 9, 1966); or 'reasonably achievable' (ICRP No. 22, 1973); or to ensure that 'no source of exposure is unjustified in relation to its benefits or those of any available alternative, that any necessary exposures are kept as low as is reasonably achievable' (ICRP No. 26, 1977). The nuclear industry today talks about dose optimisation. The problem is the basis for the 'cost-benefit' assessments seem to be performed at the national level in an informal sort of way via a kind of 'old boys' network. Such evaluations are difficult because the assessment of both costs and benefits are non-trivial and almost certainly approximate and semi-scientific. In some countries (the USA) the ICRP limits are now regarded as being 2 to 3 orders of magnitude too high.

The arguments return once again to the subjective assessment of risks. Fears might well be expressed that the dose to the public may exceed ICRP limits, perhaps because the authorities got their sums wrong or environmental surveillance was not sufficiently rigorous. Indeed it is always possible that an individual may exceed an ICRP limit; all that it means is that the risk of a subsequent deleterious event is very, very slightly higher (Penreath, 1980). The ICRP dose limits represent a level of fatality of between 10^{-5} to 10^{-6} per year, and most radiation exposures from nuclear power are kept to a small percentage of these limits. It has been suggested that risks of 10^{-3} or more per year would normally be unacceptable to society and that only risks of 10^{-4} or more per year are worthwhile spending public money on (Flowers, 1976). Risks below 10^{-5} are regarded as being generally accepted by the public without concern.

An opposite view is that of Caldicott (1980) who points out:

> The permissive radiation policy . . . turns us into guinea pigs in an experiment to determine how much radioactive material can be released into the environment before major epidemics

of cancer, leukemia, and genetic abnormalities take their toll (p. 21).

This is an extreme statement but it does reflect the fact that the safety agencies are still in a learning curve situation with respect to the health effects of low level radiation doses. It is also a comment that the current ICRP standards are predicated on the assumption that some additional radiation doses are in fact worthwhile because they allow the development of nuclear power. However, at present there seems to be a degree of latitude because the dose commitment arising from nuclear power is small. By the end of the century the margins may be much narrower. Mention could be made of the global effects of certain radionuclides such as krypton 85 which lacks any significant sinks and has a half-life of 10.7 years (Machta *et al.*, 1974). In a large nuclear programme wholebody exposures to krypton 85 may approach ICRP limits within 60 to 80 years. There may also be additional dose commitments from the reprocessing of fuel and waste storage operations.

A further aspect concerns the measurement of radiation doses. In the UK assessments are based on the concept of a critical nuclide affecting a critical group. It is recognised that individual dose rates depend on both the amount of radiation released into the environment and on its pathways through it. A comprehensive study of all possible rates of movement is very complex and an obvious simplification is to identify the one or two pathways that are overwhelmingly more important than the rest (Dunster, 1971). In this way the limits on major discharges can be set by estimating the dose rates for the few critical groups; for example, fish-eating fishermen or children drinking milk. This allows approximate estimates to be made of the 'environmental capacity' for the discharge of aqueous wastes (Preston, 1971). Such a philosophy is undoubtedly both very flexible and very convenient for the nuclear industry whilst ensuring that worst-case public doses are kept within ICRP limits. This approach has been characteristic of all nuclear discharges in the UK since the beginning of atomic energy developments. The problems arise when either pathways are missed, for example: the salt spray process of transferring plutonium from the Irish sea into a terrestrial environment; or there are no critical groups thereby allowing very high discharge limits, or if there are unexpected tendencies for radionuclides to accumulate in the environment, producing dose rates which have a lagged cumulative component to them.

In the US and in many other countries, a different approach is taken. Standards are set on a national basis and the air and water discharges are monitored. This can make it difficult for old plant to comply with new standards. Nevertheless, it would have prevented the Irish Sea in the UK from becoming the most contaminated sea in the world today. The UK response is that setting maximum standards is inevitably controversial, that they require extensive monitoring operations, and they may encourage discharge rates that are higher than need be. The answer is to have a look at the Irish sea. In 1984 15 miles of beach were closed because of the accidental release of a small amount of high-level waste. There is growing evidence that the radio-nuclides are being concentrated in the marine and estuarine environment over a large area of coastline. The levels of radiation are currently causing concern. The question can be asked, was any of this anticipated? – and the answer is 'no'. Can the environment be restored? The answer is 'no, except by natural processes'. Can the discharges be stopped? Again the answer is 'no, there is no need to because the ICRP limits are not being exceeded'. Perhaps in 30 years' time when there is a clearer perception of the medical consequences, a different view may be necessary. By then the only practicable long term remedy may involve the evacuation of certain areas.

If all else fails the emergency plan will save us

Even though nuclear plant are perfectly safe it is still considered prudent to have standard emergency plans. Apart from location, the emergency plan is the only reactor independent safety measure that is used. Current UK emergency plans cover the prospect of a limited evacuation within 2 to 3km of a reactor in the downwind sector, advice to stay indoors, the issue of potassium iodate tablets to people near the plant, and plans for the longer-term control of contaminated food and water supplies out to 40km (HSE, 1982b). The principal consequence of the Three Mile Island incident has been a review of these arrangements, resulting in the provision of additional telephone lines, a press centre, and the extension of milk monitoring out to 40km! These emergency provisions are seen as being relevant for 'reference accidents' (whatever they are). A more extensive radiation release would require far more extensive action than that envisaged in the emergency plan; if this happens it is considered that '. . . the existing emergency plans will provide the necessary immediate response in the vicinity of the site and will form the basis for actions on a wider scale' (HSE, 1982b, p. 5).

These 2 to 3km limited evacuation plans in the UK compare very unfavourably with the 16km evacuation zone now used in the US and the 80km radial zone for the control of contaminated water and food (Olds, 1981). At Three Mile Island people spontaneously evacuated themselves out to a distance of over 80km.

One problem with emergency plans is that the geographical domain must be commensurate with the possible effects of an accident. Another is that their effectiveness can only be assured by training the public through mock drills and practice evacuations. Yet at present neither are considered relevant for fear of raising questions about the safety of their friendly neighbourhood – or even regional – nuclear power plant. Yet the computer models used to predict the health consequences of accidents assume: evacuation of people within 2 hours for a 60 degree sector extending 2km downwind, 5 hours for the 2 to 5km distance band, 12 hours for the 5 to 25 km band, 1 day for the 25 to 75 km area, and 2 days for areas beyond 75 km (Hallam et al., 1982). In the US the WASH 1400 report assumed an evacuation of people in all directions to 8 km, and from the 45 degree downwind sector for 41 km.

Another problem is whether the people responsible for ordering an evacuation would respond quickly enough. The fear of bad publicity might delay such a move until it became certain that a major release had occurred. Additionally, it is often stated that in a situation where the dose seemed likely to exceed the Emergency Reference Level then countermeasures would be taken *if* a substantial reduction in dose was likely to be achieved, and *if* the countermeasures could be carried out without undue risk to the community (Marshall et al., 1983). People's judgements are involved. Problems resulting from a panic reaction might be judged far more severe than a perceived small health effect, but there would also be severe problems in predicting likely dose rates once an accident was under way.

A final problem is that the medical facilities in the region around the nuclear plant would not be prepared to handle even small numbers of contaminated people. It is doubtful whether there would be sufficient potassium iodate tablets to cover people living within more than a few kilometres of the plant. It is also doubtful whether the emergency support services – for example, environmental monitoring – would be able to respond quickly enough to provide useful advice during the management of an accident. In short, do not rely too much on the emergency plan.

The law will protect us

In the UK the Nuclear Installations Act of 1960, Section 7, states that it is a duty of the licensee (1) to secure that no occurrence involving nuclear matter causes any injury to any person or damage to any property of any person other than that of the licensee; (2) to secure that no ionising radiation emitted from anything on the site, or primary waste discharged in any form on or from the site, cause injury to any person or damage to any property other than that of the licensee. The provisions of the Act are impossible to uphold since no nuclear installation could operate without releasing some radiation, and therefore all nuclear installations must be technically in breach of section 7. Another problem is that there is no definition of 'injury'. The injury problem is in practice a difficult one because it would probably be impossible for a claimant under the act to prove liability, as the injury could be a cancer which developed many years after and only with a certain probability of having been caused by the occurrence or discharge. The operator could claim that it was not yet proven that other energy sources did not constitute similar threats and it might well be the case that the nuclear energy supply option reduced the overall cancer hazard (Stott and Taylor, 1980). In practice, this particular act offers only illusory protection. Its main contribution is to establish who is responsible for the safety of the site; and that is the site licensee.

Even if you could prove exposure to radiation – not an easy matter – then unless the doses were significant in terms of the ICRP recommendations, virtually no one will believe that there was any real health effect attributable to it. This leads to an interesting dilemma. There are two very different ways of proving that nuclear power is safe. One way is to declare that it is safe and harmless unless it can be proven otherwise. Thus it can be declared that clusters of leukaemias around nuclear facilities could not possibly be linked to radiation exposures because the evidence is purely circumstantial; the doses are too small and the causal links have not been identified. This 'safe until proven otherwise' approach employs what is termed 'the optimistic null hypothesis of no significant effects'. It is widely used in the UK when dealing with potentially hazardous substances. Until large numbers of people are killed in a clearly culpable fashion they are assumed safe, by default.

The other approach is to put the burden of proof on those who believe that nuclear power is safe. Let them demonstrate in clear and convincing manner that any circumstantial health effects are not related to radiation releases. The choice of hypothesis that nuclear power is either safe or unsafe is essentially arbitrary and at present neither can be proven or disproven. The choice is largely a matter of faith rather than science.

Conclusions

We are heading towards an asymptotic all-nuclear future. Public acceptability and public safety both need to be preserved although the former is more difficult than the latter. Reactors not only have to be safe but they must be seen to be safe and the same applies to the entire fuel cycle. As Cottrell (1981) put it:

> One or two more accidents like that at Three Mile Island, coming fairly soon after one another in well populated regions, and even with no casualties, would surely make nuclear power totally unacceptable to the general public (p. 70).

To which it may be added that even if it were to become unacceptable, it may still be required because there could well be no alternative source of energy. It is clear then that the isolation of nuclear plants in sparsely populated regions may well be the necessary price for the long term public acceptance of nuclear power. If nothing else could be done, then the public may well be prepared to accept it as a necessary evil. Certainly, large-scale propaganda exercises will by themselves never succeed if the unthinkable happens and rational debate relies heavily on the excellent safety record continuing. As Mathews and Usher (1977) say: '. . . it is the practical demonstration of the excellent safety of the CEGB's nuclear stations since they commenced operation in 1962 that has been a major factor in public acceptance' (p. 154). These sorts of public assurance are already looking a little jaded.

Important problems concern the increasing need to prove that a reactor system has been built according to design. The CEGB's criteria that all accidents leading to an uncontrolled release of radiation should be less than 10^{-6} per reactor year is only a design target (Matthews, 1982). However, even then, the fact a reactor has been built according to design provides no assurance that it is safe. It is often assumed that the plant is adequately designed and susceptible only to random malfunctions. The fact that safety systems may not always function perfectly even when its component parts operate perfectly and with no operator error, needs to be considered. In the past this has applied to the emergency core cooling systems used on PWRs but it could also apply to other safety systems. Another problem is the necessary assumption that operators do not make errors and that management is competent. The UK perception of Three Mile Island is that it was due to operator error. The suggested solution is to use a computer controlled automatic shutdown procedure with no need for operator intervention. One imagines this to be a possible recipe for disaster: it only requires a small degree of software unreliability, perhaps combined with sensor failures, to initiate the first reactor accident attributable to computer error. The control systems have to be perfect and it is not at all certain how this may be achieved. Computerisation does not remove the very real risks of operator errors which have to be completely eliminated if the reactor design specification is to be regarded as a relevant indication of reactor safety; and this task is impossible.

A further difficulty concerns the prospect of sabotage. The UCS (1977) concludes that the risks of reactor accidents due to sabotage are probably of an order of magnitude greater than any other cause. It is sometimes claimed that nuclear plants are difficult to sabotage successfully, and that accordingly they are far less susceptible than other potential targets. On the other hand the resulting publicity could well make nuclear plant very 'attractive' to terrorist organisations. It should also be noted that the consequences of a successful attack might well be far worse than the worst predicted by accident studies, mainly because the low likelihood of a normal accident is accompanied by a high likelihood of mitigating circumstances. The point here is that whilst it may well be physically difficult to sabotage a reactor, it would not be impossible, especially if the terrorists had the necessary skills with explosives and reactor engineering. There are innumerable ways of disrupting cooling water, disrupting power supplies, damaging the containment and the spent fuel ponds. If a plant cannot be made completely secure then consideration should be given to remote siting and the phasing out of the most attractive urban plants which have the highest potential terrorist risk factors.

None of this discussion is meant to imply that nuclear plants are necessarily unsafe or that their designs are flawed, but merely to point out that at present it is impossible to demonstrate that the risks are zero or even what the real risks and their associated error bands are (El-Hinnawi, 1980). It will be well into the next century before there can be sufficient empirical evidence about the spectrum of accident frequencies. If the best sites from a public safety point of view are not used, then should a major accident ever occur, the public may well be surprised – and not without reason – that their safety should have been sacrificed for economic or aesthetic considerations, and that more attention had not been paid to the use of geographical isolation as an additional safety factor (Openshaw, 1982). The use of isolated sites that are not as near major load centres as the power utilities may have wished carries with it economic and environmental costs; but perhaps these are the price of the insurance premium we have to pay for wanting nuclear power. The more densely populated the country, the greater the environmental conflict, and the greater the need for isolated sites.

The nuclear industry wants a balance between safety and economic considerations. It can be argued that they are responsible for making nuclear power work and it must be their decision. However, it does seem that even people who are generally willing to accept the idea that safety cannot be absolute seem to feel that nuclear power is an exception. Perhaps the best compromise for both sides is to utilise geographical isolation as an additional safety factor. The geography of risk is strongly influenced by the choice of sites and sites could be selected to minimise total societal risks. There is nothing new in this suggestion. In the UK it is already a well established safety principle that 'All reasonably practical steps shall be taken to minimise the radiological consequences of any accident' (HSE, 1979, p. 3). Site selection is one way of meeting this goal. In the US there seems to be a greater acceptance of the need for geographical isolation, probably due in no small part to the repeal of the limited liability provisions of the Price Anderson Act in 1982. Utilities are now financially liable for the consequences of any accidents and they can minimise these risks by seeking remote sites. The CEGB's £5 million limited liability does not offer anything like a similar incentive. In the inevitable cost benefit assessments £5 million does not buy many miles of power line.

So it can be concluded that a risk-adverse public and a risk-tolerant nuclear industry could well find their best long-term interests served by remote siting. The former for psychological and health reasons, distance mitigates most consequences to some degree. If you do not trust what 'they say', if you do not believe 'their' assurances of safety, and if you suspect that the 'approved' safe discharge limits are in fact unsafe, then geographical isolation may be an effective palliative. If the nuclear industry fears bad publicity, if they believe the future is uncertain because of the possibility that major accidents will occur somewhere in the world, and if nuclear power is an absolute necessity for the future of man, then a risk-adverse nuclear industry would seek to anticipate possible demands for the closure of 'dangerous' sites next to areas of high population density by seeking suitably remote sites for the next generation of

43

nuclear power plant. Any other course of action would smack of a distinct gamble with public safety and future power supplies.

2 Power station planning in the UK

Introduction

The purpose of this chapter is to document and explain the nuclear power station planning process as it is practised in the UK, in an attempt to find out why isolated sites have never been considered appropriate. In some senses it is a sociological study which tries to understand how the system works and to offer a justification for it. Subsequently it is placed within a broader framework and a number of important criticisms of current practices are made.

First, then, some background information about the quasi-governmental agencies who are responsible for siting and operating nuclear power stations. The Electricity Act of 1947 nationalised the electricity supply and replaced the previous mixture of public and private utilities. The Electricity Act of 1957 created the present administrative structure. In England and Wales, the Central Electricity Generating Board (CEGB) is responsible for generating or acquiring bulk supplies of electricity. In Scotland, the equivalent body is the South of Scotland Electricity Board (SSEB). In England the CEGB supplies bulk electricity to 12 Area Boards who in turn distribute it to consumers. A third body, the Electricity Council, is responsible for research, development, labour relations, and general policy advice. The CEGB, Area Boards, and the Electricity Council report annually to the Secretary of State for Energy who in turn reports to Parliament. Planning control is effected by mandatory ministerial consents to certain types of development. For power station construction or extension, consent is required under the Electric Lighting Act of 1909 Section 2. Table 2.1 shows the other legislation under which consents are needed to build and operate a power station. Finally financial approval has to be obtained from the Treasury.

45

TABLE 2.1 *Legislation covering the development of a nuclear power station*

Act of Parliament	Purpose
Electric Lighting Act, 1909	Consent to build a power station
Town & Country Planning Act, 1971	Details of buildings and land use
Nuclear Installations Act, 1963	A site licence for a nuclear reactor
Radioactive Substances Act, 1960	Storage and transport of radioactive material
Dumping at Sea Act, 1974	Dumping of substances below high water mark
Coast Protection Act, 1949	Consent for works below high water mark
Water Resources Act, 1963	Consent for the abstraction of cooling water
Rivers Prevention of Pollution Act, 1951	Consent for the discharge of cooling water and trade effluent
Land Drainage (Amendment) Act, 1976	Consent for changes in land drainage
Control of Pollution Act, 1974	Licence for waste disposal, control of noise, and other matters

Statutory duties

Section 2 of the 1957 Electricity Act that created the CEGB states that the Board's main duty is

> . . . to develop and maintain an efficient, co-ordinated and economical system of supply of electricity in bulk for all parts of England and Wales.

The South of Scotland Electricity Board (SSEB) has a similar statutory responsibility for Scotland. That is to say the Boards *have to meet* consumer needs on a continuous basis subject to the provisions of section 37 of the 1957 Act, which states that:

> In formulating or considering any proposals relating to the functions of the Generating Board or any of the Area Boards . . . , the Board in question, the Electricity Council and the Minister, having regard to the desirability of preserving natural beauty, of conserving flora, fauna, and geological or physiographical features of special interest, and of protecting buildings and other objects of architectural or historical interest, shall take into account any effect which the proposals

would have on the natural beauty of the countryside or on any such flora, fauna, features, buildings or objects.

This statutory duty applies to all activities of the Board from transmission lines to nuclear power stations. It is a very onerous duty because of the need to strike a balance between the inevitable changes that power developments bring, various technical and economic costs, and the benefits to the consumer. It is not easy because of the difficulty of evaluating the various components using different objective and subjective measurement scales. It is also not easy to strike a good balance, and precisely how the Board achieves this feat is not at all clear. The outcome seems to be a result of informal debate among Board members based on personal and expert judgements and opinions of those involved. Whether or not the introduction of formal evaluation methods would improve this process is commented upon later.

In seeking this balance between amenity and the provision of power, the Board puts a high degree of emphasis on its national duty and on national interest arguments. They have a statutory responsibility to provide a national good in the form of a continuous supply of electricity and, whilst amenity is a constraint and local factors may be taken into account, in the end unpopular decisions may have to be made. The Board view themselves as being left with many such very onerous decisions which have to be made in the national interest. The Board seem to view the process as one of extremes; either you accept the proposed developments or you face power cuts. Opponents to nuclear power are either regarded as misguided or in favour of a different, non-industrial life style (Matthews and Usher, 1977).

The planning problems are complicated by the long lead times associated with power station developments. Planners have to work 12 to 14 years ahead of demand. Assuming an adopted site exists, then planning takes about 4 years. This time is seen as a result of the thoroughness by which the various engineering and environmental surveys are performed and the need to incorporate democratic procedures and various extensive consultations into the planning process. The construction of a new nuclear power station will probably take at least 5 years and probably 8 years from the moment the decision to initiate the process is made. Before any work can start it is necessary to receive the Secretary of State's consent together with a number of related consents and licences, in addition to financial approval from Government; and this may add a few more years.

The CEGB operate an annual planning cycle and include

provision for specific new stations that are required for commissioning in the period 7 to 9 years ahead. To do this it is necessary to secure adequate sites for up to 12 years ahead, otherwise future development programmes could be delayed resulting in (as they see it) power cuts, voltage reductions, and insufficient power to meet national economic and domestic needs. This makes it necessary to forecast power demands for 12 to 20 years ahead in order to allow sufficient time for new facilities to be built. In addition to the need for new plant, Section 2 of the 1957 Act is now being interpreted as providing a duty to offer greater economy to the consumer; industrial consumers have an effective lobby to reduce electricity prices. It seems that increasing emphasis is now being placed on the 1978-79 corporate objective,

> To maintain and develop supplies of electricity to meet the needs of customers in England and Wales on a continuing basis as cheaply as possible (Electricity Council, 1983, p. 11).

The 1957 Act uses the term 'economical' and this was originally interpreted as not necessarily implying 'cheapest possible' because of environmental and security of supply factors. However, a major justification for the move towards nuclear power is this economic argument; nuclear power is viewed as being cheaper than any other form of generation despite considerable criticisms of the economic assumptions.

The 1957 Act is also interpreted as requiring actions to improve the security of fuel supplies by diversifying the type of fuel used. At present about 86 per cent of power generation is coal fired. This may well be a political constraint aimed at reducing the political leverage of striking coal miners; but it is also common sense to diversify the type of fuel used. The arguments seem to have been extended even further than this. It is being suggested at the Sizewell B public inquiry that diversification towards nuclear power may also be justified in terms of possible future benefits, even if there is no increase in demand. The estimated savings in fuel costs are assumed to exceed the life time costs of the PWR. Visions of export markets and possible future legislation regarding SO^2 emissions from fossil fuelled stations are also involved. It is worth emphasising that the CEGB does not have a will of its own but answers to Parliament, and that it has to interpret governmental wishes whilst seeking to discharge its statutory functions. Precisely what these influences are, and how they operate, is not public knowledge. There is nothing sinister about the CEGB; they are, in many ways, a typical central government agency which in a

UK-style democracy is answerable to Parliament and not directly to the public.

The needs element

The requirement for accurate, long term forecasts of power demand is very difficult to achieve in practice. Historical extrapolations of past trends have generally failed to provide the basis for accurate forecasts. Prior to 1962 the industry generally underestimated demand as a result of underestimating the national economic growth rate, the rapid utilisation of space heaters, and mistaken assumptions about system load factors (Reid *et al.*, 1973). Since 1962, demand has been consistently overestimated as a result of a slower than expected national economic growth and competition from natural gas. The resulting overcapacity has presented the industry with difficult management and public relations problems. It has also greatly slowed the use of nuclear power. At present there is a considerable oversupply of electricity. Some of the excess has been used as an additional planning margin to allow for breakdowns, system failure, and exceptionally bad winters; some to allow for replacement of old, uneconomic but not yet obsolete or unserviceable, plant. The planning margin at present stands at 28 per cent (it would have been even larger if AGRs ordered in the 1960s were operating), and this is sufficient to meet the winter peak half-hour in full except for 19 winters per century; and it should avoid the need for disconnections in all but 4 winters per century. The previous 1964 generating security standard specified a target of 23 and 3 respectively.

The CEGB needs to consider likely patterns of demand for the next 20 years with some thoughts for the next 50 years up to 2030 AD. They do this via a scenario-based approach for the UK economy. One problem this type of approach is supposed to overcome occurs if there is low growth in the short term (whilst current economic problems are being overcome) and much faster growth in the long term (when the economic problems have been overcome); if this occurs, new economic problems might result from shortages of electricity supply.

The logic of a scenario based approach is described in Hutber (1984). The CEGB outline 5 scenarios in their Statement of Case for the Sizewell B public inquiry (CEGB, 1982, Appendix D). Scenario A involves high growth based on services; scenario B high growth based on manufacturing similar to the period 1950 to 1960; scenario C medium economic growth with a mixture of success and failure similar to the last decade; scenario D stable

49

low economic growth with de-industrialisation and a declining level of economic activity; and scenario E with unstable low economic growth, unplanned deindustrialisation and an inability to overcome economic problems. The idea is to examine important decisions against alternative futures and to offer a good indication of the credible range of possible outcomes for planning purposes. The range has to be sufficiently wide to form a sound basis for planning so that plans are sufficiently flexible and robust to meet whatever requirements the scenarios show up. The results in terms of maximum electricity demand for 2000 AD are given in Table 2.2.

TABLE 2.2 *Forecasts of electricity demand for England and Wales in 2000 AD*

| | Demand growth scenarios | | | | | |
	A	B	C	D	E	1979/80 actual
Maximum demand (GW e)	51.3	61.7	46.9	35.1	35.1	44.1
Annual GDP growth	2.6	2.6	1.0	-.4	-.4	

Source: CEGB (1982), p. 225, 235.

The range of demand of 35 to 62GW is too large for practical planning purposes. So scenario C has been selected as being the most appropriate, even though it might be argued that D is more realistic. In any case it is considered likely that the siting requirements for scenario C will also satisfy the only other likely scenarios (B and D). For purposes of economic evaluation it is envisaged that the forecast demand for scenario C can be met in three ways: a no-nuclear option, a medium nuclear option leading to 40 per cent nuclear capacity (the OECD norm for 2000 AD, see OECD (1977)), and a high nuclear option leading to 70 per cent nuclear capacity. The numbers of power stations needed for each background are shown in Table 2.3.

At present and up until 1992-3 no new capacity is needed. But under scenario C some 9GW will be needed by 2000 AD. If this is to be met by the most economic means available – by PWRs – then it would be necessary to start ordering the new plants sufficiently early to allow an adequate rate of building of the new plant in time to meet the cumulative requirements by 2000 AD. By contrast, if scenario B was to be met then some 28GW of new capacity would be required by 2000 AD. If on the other hand scenario E occurs, then whilst no new capacity would be needed some would be required to replace old existing plant in an

50

TABLE 2.3 *Numbers of new power stations needed to meet scenario C (in 1GW e units)*

	Capacity in 1981	no new nuclear by 2000	2030	medium nuclear by 2000	2030	high nuclear background by 2000	2030
Coal fired	38	8	59	1	34	0	13
Nuclear	4.5	0	0	9	30	12	54
Oil	10.6	0	0	0	0	0	0

Source: CEGB (1982), p. 55.

orderly manner. It is estimated that by 2020 AD nearly all the existing coal fired plant will have to be replaced. So it seems that a combination of even a modest level of economic growth (say 1 per cent per annum) and the retirement of old plant either on engineering or economic grounds, might easily require one new station per year from the early 1990s onwards. These stations will have to be ordered fairly soon to avoid possible electricity shortages in the early 2000s and sites found now or reserved for these developments.

Some indication of the numbers of sites needed is given in Table 2.3; it may be assumed that each GW is equivalent to one station. The nature or type of site depends on what sort of plant is expected. The CEGB is very flexible (some might say grossly inefficient) in that it holds options for sites which it may never take up. Additionally, there is no direct link between the search for sites and the estimates of future demand. The CEGB's planning department is continually searching for sites in rough anticipation of possible future demands. Thus whenever a demand requires a new station in a particular region there is an inventory of potential sites from which one can be chosen. An indication of possible sites has been leaked from the CEGB and published by the Suffolk Preservation Society (SPS, 1982). This list was apparently prepared in the late 1970s for the Watt Committee on Energy report published in 1979. Figure 2.1 shows the locations of these sites plus others that the CEGB are known to be interested in. It would appear that most of the coastline has been examined at one time or another.

The number of sites in Figure 2.1 fit in well with the scenario C assumptions for the medium nuclear and high nuclear backgrounds. There are, of course, no commitments to any of these sites apart from Sizewell B for which a planning application has

51

100 Km

FIGURE 2.1 Map of possible nuclear power station sites in Britain

been submitted. The Board would clearly like to build up a pool of planning consents independent of the actual perceived demands so that the lead time needed for planning can be reduced. In the early 1970s this process was applied to Sizewell and Torness in Scotland. Consent was also given for an oil fired station at Insworke Point in 1973, even though it is now most unlikely that any more oil fired plant will ever again be built. The same applies to some extent to coal fired options. Whether any coal fired options are taken up probably depends on the performance of the existing AGR and future PWR stations; if they work reliably then it is currently thought unlikely that any more coal fired plant will ever again be built. This is not a new situation but has been characteristic of CEGB thinking for at least a decade. The last coal station (Drax B) was only built after the government agreed to compensate the CEGB for the additional costs of burning coal. It was not a rational economic decision, but then current nuclear intentions may also be irrational (Sweet, 1983; Jeffery, 1982).

Finding a site for a nuclear station

The background energy demand scenarios provide a basis for the more detailed site searches. The planning process that leads to the construction of a nuclear station has evolved over many years and involves four broad stages:
(1) the area of search phase;
(2) the detailed site investigation phase;
(3) the consents phase; and
(4) the design and construction phase.
Table 2.4 provides a summary of what is involved. It would seem that, with the passage of time and the accumulation of knowledge, the first phase has become more a review of known sites rather than a continual search for new sites. Figure 2.2 provides a visual representation of how the CEGB view the planning process.

The area of search phase

In terms of having to meet a national, and perhaps a local need, attention is usually focused on a particular Area Board's region. For example, there may be a need for more power in the London area and this may best be met on a national basis by providing additional capacity in the north and exporting power southwards, as well as by the construction of new plants in the south-east and south-west. The current intention to reduce oil-burn means that

53

TABLE 2.4 *Stages in the planning of a nuclear power station*

Stage	Activities
1 Area of search	• identify area with a need for a new development • review existing information • determine technical and environmental potential • seek views from Government Departments, Countryside Commission, Nature Conservancy Council, and local planning authorities • identify a short list of sites worthy of investigation
2 Detailed investigation	• public announcement that certain sites are of possible interest • detailed investigations to determine technical suitability • studies of impact and amenity • detailed discussions with Government Departments, local authorities, and other statutory authorities • views invited on the merits and impacts of developments and where further research is needed • reactions of individuals, societies, and local authorities etc received • relative merits of alternatives sites assessed • balance drawn between technical/economic and amenity factors • site adopted and purchased
3 Selection of site and application for consent	• application for consent made when needs arise • Form B submitted to local planning authority to obtain formal views • notice of application published • Public Inquiry may be required before planning approval
4 Design and construction	• financial sanction • site preparation works • construction period

Source: Howells and Gammon (1982).

FIGURE 2.2 CEGB view of the power station planning process

replacement power must be imported from the north into the south where most of the oil fired capacity is located. In national terms it may be sensible to build nuclear stations in both the north and the south at the same time. This is because the development of the national grid and super-grid networks as a national bulk electricity distribution system means that the best locations in terms of the transmission system no longer need be located adjacent to major load centres only. It also means that local demand forecasting may no longer determine the location of new capacity; although it is a good public relations measure to pretend that it does.

The CEGB's list of possible sites can be used as the basis for a more detailed study leading to the adoption of a particular site as being technically suitable for a nuclear station. The selection process seems to start with an informal survey of the available sites plus other potential sites. These surveys are performed with a set of criteria in mind (see next section), and possible local objections may be given consideration. As Hunt (1970) puts it

> The search for a site at a particular part of the country starts with a study of all published information. The choice can usually be narrowed down fairly quickly to a number of specific localities which must be studied in detail (p. 5).

The early stations were located in coal deficient regions in order to avoid the cost of coal transport or the long distance transmission of electricity. The 400kV supergrid had not yet been built and the shape of the 132kV national grid could be adjusted to provide short connections to the proposed sites. Now there are over 8000 km of 400kV lines and detailed studies of power flow forecasts, system security, and transmission economics provide broad indications of where to site new stations to make best use of the supergrid. Gammon and Pedgrift (1983) note that

Increasingly such considerations have led the search for sites into regions where generation capacity will not match electricity demands, like the north-east and the south-west of England. Within such regions, assessments are made of alternative sites and their respective transmission connections to the grid. These cover the full range of economic considerations; the capital and operating costs of the station, the effect on the rest of the national power system, operation costs and the costs of new lines and transmission equipment (pp. 44-5).

It should be noted that the 400kV system favours a dispersed pattern of power generation facilities to cover the base load situation. Adams and Stone (1967) pointed out:

In the next decade stations will mainly be required in the south and northwest, but energy economics will also dictate a progressive spread (of nuclear power) to the central and north-eastern coal field areas (p. 30).

The detailed site investigation phase

Once one or more sites have been short listed, consultations start with the local authorities, land owners, other relevant bodies, and with central government departments who may have a vested interest in a site. This is necessary in order to hear their views and determine what the site is really like. At the same time a number of detailed technical studies are performed including the following: boreholes to examine the subsoil structure and geology; seismicity, hydrogeological, marine surveys of currents and thermal diffusion; coastal stability studies; an examination of flood levels based on the notional 1 in 100 years flood at the highest astronomical tide; identification of external hazards; suitability of fresh water supplies; the provision of transport facilities particularly a site for a road-rail transhipment point; and an accurate population count to verify that any emergency plans could in fact be operationalised. Account is also taken of any long-term development plans for either large-scale residential expansion or for industrial developments which would significantly change the characteristics of the site. If all is well then a decision may be made to adopt the site. This can be a difficult decision and there may be no easy answer to problems resulting from the conflict between different criteria and the presence of local opposition. At the end of the day a decision must be made. National necessities require a new plant and this may make it inevitable that some individuals and organisations will be offended to some degree. It is usually claimed that there is very little flexibility in site selection because of the siting constraints that have to be satisfied. At this point political factors may enter into consideration. The Board must take account of political influences and the implications for siting nuclear plant. The CEGB characterise their views as follows:

> The CEGB are not influenced by the *fact* of opposition but by the *merits* of the points made by opponents. The Secretary of State takes political considerations into account in arriving at his decision to grant or refuse consent (Drapkin, 1974, p. 225, based on letter from CEGB).

In short unless there are good technical reasons for refusing permission, development is likely to proceed.

The consents phase

Once an adoption decision is made, the selection is announced in public. There is no commitment to actually develop the site as yet, all the Board will declare is that it has adopted a particular location as a potential nuclear power station site and that it intends to submit an application for consent when the need requires it. This causes a degree of planning blight and the Board will rarely go to this stage without a firm intention for an early development. It may also try to purchase the site on the open market if the opportunity arises. The fact that there may be strong local opposition to the adoption of the site is not sufficient to cause the site to be rejected since it is currently thought that all new greenfield sites, and indeed existing ones, will be opposed by people who perhaps fail to appreciate the significance and importance nationally of the Board's proposals and work. The selection process at this point is still internal to the Board.

The Board will now undertake extensive pre-application discussions. Formal consultations do not start until after the application is made, but the Board will not want to make an application for consent if it is likely to fail. So they will negotiate, discuss, modify, suggest, and consider alternatives to the details of their proposal with government departments, planning and other authorities, and with other statutory bodies. One objective is to try and remove possible objections to a formal application by negotiation and explanation of why the site is needed. Once the application is made the Board will have taken its decision in principle. The only real chance of influencing the Board, and it's a very small one, is at this pre-application stage when the Board is not publicly committed to a specific site; the commitment is made by applying for planning consent at which time considerable detailed design work will have been done.

An application may eventually be made under the Electric Lighting Act of 1909. Section 2 of this act requires the Board to apply to the Secretary of State for Energy for consent to build a power station. According to Section 40 of the Town and Country Planning Act of 1971, the minister's consent may carry with it deemed planning permission. That is, the Board as a statutory undertaker is not required to apply to local authorities for planning permission. If the local planning authority objects to the application then there must be a public inquiry. If there is no local authority objection, the decision on whether or not to hold a public inquiry rests with the Department of Energy and is discretionary. The Sizewell B public inquiry was called by the Department of Energy because of a promise that there would be an extensive public inquiry prior to building the first PWR reactor in the UK. It may eventually be seen as a generic inquiry that removes the need for any subsequent inquiries to cover the basic safety and policy issues; this may well require modification to the 1971 Town and Country Planning Act.

The purpose of the public inquiry is to inform the Minister so that he may have sufficient information to make the best decision. As Garner (1970) puts it

> An inquiry . . . is essentially a fact-finding agent, convened to ascertain what the facts may be relating to the matters in issue, with the object of so informing the Minister . . . in order that he may better be able to come to a decision on the questions before him (p. 93).

It is not charged with the decision-making, a fact often not properly appreciated by opponents of the Board's schemes, and people's expectations can be unreasonably inflated by a public inquiry. The inspectors, an electricity inspector, a Department of Environment co-inspector, and perhaps an assessor, prepare a report and make recommendations. The Electricity Inspectorate dates back to the Electric Lighting Act of 1899. Their concern is with safety, security of supply, and impact on amenity. Generally, they would have engineering backgrounds. The

Department of Environment inspector will usually have a background in planning, surveying, law, architecture or engineering. The assessor would normally deal with either emissions or health effects.

In most instances the Board would naturally expect the inspector's report to recommend the Board's proposal to the Minister. The Minister will almost invariably accept these recommendations but the report is only meant to be a basis for the decision. The Department of Energy also has an influence on the final decision, and eventually a position will be recommended to the Minister and a final decision made. There is no right of appeal.

The CEGB do not always have it their own way, although usually they do – a fact they would attribute to the thoroughness of their extensive preparations before applying for consent, including their discussion and negotiations with most of the parties in a position to raise any formidable opposition. It is in any case' unlikely that the Board would make an application unless they were very certain of their chances of success. After all, the consents process is very time-consuming and expensive, so most of the major issues and many of the details are resolved before a public inquiry is convened. The Board also view their application as a reflection of governmental policy and would expect some support from the Minister. However, either the inspector or the minister can impose conditions on the consent.

In many ways this inquiry system is unfair to the opponents and it can be very slow in reaching a conclusion. There are ways of streamlining it and these would probably be necessary if there were to be a large number of public inquiries; for example, resolution of all the generic issues in advance, restricting the terms of reference to local issues, and changing the law so that fewer inquiries would be required. In general it seems that the Board view it as a nuisance that can cause expensive delays and costs large sums of public money; the opponents receive no public funding; but it also acts as a safety valve for reducing local tension by letting the opponents air their views and question the Board's witnesses in public.

The design and construction phase

Once approval is obtained, work commences. However there are still opportunities for minor modifications to landscaping, facia colouring, access roads etc., and some consultations continue until construction is complete. The application of this process to particular sites is considered in some detail in a later chapter.

59

Criteria for selecting sites

The criteria used to screen sites represent the most interesting geographical aspects of the whole process, and it is seemingly a particular area where geographers with an interest in locational analysis might be able to apply their expertise. It is also a major area where present procedures can be questioned.

TABLE 2.5 *List of technical siting criteria*

1 An abundant and unfailing supply of water for condenser cooling
2 Foundation conditions capable of supporting heavy loads under all conditions
3 An adequate area of reasonably flat land at an acceptable height above the source of cooling water
4 A location strategically placed for connection to the transmission grid and so positioned to supply electricity to areas of high demand
5 Access for construction
6 The nature of nearby industries and other sources of external hazard must present no threat to nuclear safety
7 Nuclear safety
8 Effects on local communities
9 Acceptable environmental effects
10 A question of balance

In theory at least, nuclear power stations have the unique advantage that the cost of producing a unit of electricity is broadly the same irrespective of where the station is located. This is of course very different from coal and oil fired plant where proximity to coal supplies and oil refineries are major geographical constraints. However, whilst nuclear plant appears to be 'foot-loose' there are other criteria that restrict the siting process. Some of these factors are physical, others economic and relate to the operation of the national grid system. There are a number of technical requirements for a power station site and these are listed in Table 2.5. Each is discussed in turn (see Hunt (1970) and Haire and Usher (1975) for further details).

1 An abundant and unfailing supply of water for condenser cooling

This requirement for large quantities of cooling water is probably the primary factor in siting nuclear power stations. The low steam pressures of the early reactors (MAGNOX stations) meant that in general, only coastal or estuarine sites (on the Seven, Trent, and Thames) would be able to satisfy water demands if direct cooling was used. Most rivers in the UK have too small a flow and would require expensive dry cooling towers, but even then the summer flow rates may be insufficient. The situation with respect to the later AGR stations is more flexible, the higher steam temperatures resulting in a reduced demand for water. Nevertheless, the amounts of water required are considerable; for example, it is estimated that the Sizewell B PWR will need about 55 cubic metres per second plus 0.5 million gallons of potable water.

The coastal sites also require extensive investigation to ensure that the warm reject water diffuses rapidly enough to avoid the possibility of re-entry, and that the thermal plume is dispersed without any harmful effects on the aquatic ecology. These are important details because the reject water may easily be 10°C above ambient sea temperatures. It is usually stated that the localised warming of the sea is good for marine life, although the effects of non-continuity of the warm water is not so good for fish.

The main argument against the use of coastal sites is the difficulty of finding and securing sites, and the associated problems of obtaining easements for the transmission lines necessary to convey the power to inland load centres. Hunt (1970) explains that there is accordingly '. . . a strong incentive for developing nuclear stations inland in the future' (p. 651). This would probably require groups of cooling towers of some 114 metres in height. Hunt is concerned, wrongly as it turns out, that

> Current experience demonstrates that the visual and
> environmental effects of cooling towers promises in the near
> future to be as much a subject of public debate as the safety
> considerations of near urban siting (p. 651).

2 Foundation conditions capable of supporting heavy loads under all conditions

Good foundations are essential to support the weight of a reactor which may amount to 100,000 tons. The critical factor is the need to avoid differential settlement. Additionally, known fault-lines are avoided even if they are not geologically active and micro seismicity studies are carried out. Hunt (1970) considers that

> Good foundation conditions – preferably rock a few feet below
> ground level – are important as one of the main factors
> affecting the relative economies of sites (p. 652).

This requirement, however, has not prevented sites being selected where there is no solid rock a few feet below the ground;

indeed only two of the MAGNOX stations are on sites with firm rock foundations, three are located on soft marl, one on shingle, and one on sand (Hinton, 1963). Subsurface geology mainly influences site development costs whilst the presence of old mining tunnels may easily preclude a site from being developed.

3 An adequate area of reasonably flat land at an acceptable height above the source of cooling water

Flat and level site areas of 40 to 80 ha are necessary for nuclear plant, although in practice sites of at least 80 ha would be sought in order to provide development potential, that is multiple reactors on the same site. Sites should be above flood level (10 to 15 metres above high water mark) and protected from surface flooding. However to reduce pumping costs the total head of water must be kept low by maintaining syphonic recovery at all but extreme low tides. On high land this would mean deep excavations for the turbine hall. For coastal sites these factors imply costly site works for site levels outside the range 3 to 15 metres above sea level. The constraints are again largely economic, and would imply that cliff sites would not normally be considered – but they are not impossible either.

4 A location strategically placed for connection to the transmission grid and so positioned to supply electricity to areas of high demand

This is another economic constraint. Suitable routes must be available for connections to the super grid. Indeed proximity to the super grid system is a very important locational constraint. As Gammon and Pedgrift (1983) emphasise:

> Now the 400kV supergrid system is largely completed, system considerations and the desirability of avoiding long new transmission lines have become important factors in choosing where new stations should be located (p. 41).

The reasons are primarily cost, objections on amenity grounds to new line developments (although these may be needed subsequently anyway), and convenience. It also implies that sites will increasingly be sought in areas where there is unused transmission line capacity. Thus it seems that the existing national system of electricity generation, plus forecast changes in demand, will generally indicate those areas of the country in which to locate new plants; and this factor, perhaps more than any other, is the key component in locating nuclear power stations.

5 Access for construction

New roads may have to be provided to deal with the large number of trucks transporting construction materials, especially sand and aggregates, and to reduce nuisance levels to local people. It is estimated that sand and gravel amounting to about 650,000 tonnes and 140,000 tonnes of cement would be needed for the construction of Sizewell B; or about 50 road vehicles each of 20 tonnes daily for 3 years. In addition, there would be about 5 vehicles per day carrying pulverised fuel ash, and 4 carrying steel. During times of peak activity there could be 200 loaded vehicle movements into and out of the site per day with perhaps some 160,000 deliveries during construction. Provision would also have to be made for the transport of a small number of heavy loads; the reactor pressure vessel (420 tonnes) and the steam generators (330 tonnes each). Sea transport would probably be used. Other loads may be lighter but very large in size. This mainly affects site specific development costs, since the CEGB may have to pay for the road improvements and face angry residents upset by the large numbers of lorry movements. It also means that inaccessible sites – for example, at the foot of a high cliff – would not normally be selected because of the problems of access and the relatively high site preparation costs that would arise.

6 The nature of nearby industries and other sources of external hazard must present no threat to nuclear safety

Until recently this meant avoiding airfields. Now it seems that aircraft are regarded as presenting a greater hazard to nuclear plant than previously recognised. The RAF apparently instruct their pilots to regard nuclear stations as being enemy missile batteries, although the USAF sometimes use them for navigation purposes. The proximity of chemical storage tanks may *now* result in sites being avoided, although it seems that the importance of this sort of hazard (toxic gas or gas cloud explosions) is something that has recently grown in importance. It is also something that development control restrictions have failed to prevent. This is a reflection of the belief that the safety distances for chemical storage tanks are very small and that no special precautions need be taken at the nearby nuclear stations. There are indications that a more realistic view is being adopted for new plant, the attitude being that whilst the nuclear plant is completely safe the other industries may not be, and could affect nuclear safety. Considerations such as these become far more important if industrial uses are being sought for nuclear power and if urban siting is being sought.

63

7 Nuclear safety

As Hunt (1970) puts it '. . . the site must comply with such local population standards as the licensing authority may lay down' (p. 653). This factor ranks last in the technical factors listed by Hunt, although too much importance should not be attached to his ordering. In practice the population constraints are the only criteria that are explicitly defined. They are considered in detail in Chapters 3 and 4. The population limits used are determined by the Nuclear Installation Inspectorate (HSE, 1982) and not the CEGB.

8 Effects on local communities

The CEGB have recently been investigating the effects that power-station building and operating activities have on local communities in the areas nearby. A series of research reports by the Power Station Impacts Research Team of Oxford Polytechnic, funded by the CEGB, have investigated various local social and economic effects (PSI, 1982). There is no indication whatsoever that the CEGB have ever taken such factors into account when selecting sites or that they intend to in the future, but it is information the Board need to have at their disposal. It would seem to be useful for two purposes: to persuade local communities that (nuclear) power stations are 'good' developments in terms of their effects on the local economy, or if not 'good' are at least neutral in their effects; and, perhaps, to show that power stations are not as 'bad' as people might have believed in remote areas.

9 Acceptable environmental effects

The 1957 Electricity Act requires that environmental and economic considerations are of equal importance. Obviously any power station development will damage the environment both physically and aesthetically (only a few people would disagree). Sites have *generally* been sought to avoid National Parks; green belts; areas of outstanding national beauty; areas of great landscape, historical, or scientific value; and areas of unique aquatic or ecological interest. There are exceptions and one important factor may be the extent to which the features being damaged are unique. This aspect of siting is by no means easy and it is discussed at some length later with respect to Sizewell and two other case study sites. It is perhaps the most controversial of the criteria. the CEGB seem to follow, if they can, a line of least public and planning opposition. As Hunt (1970) writes:

> It is not easy to find sites for major power stations which meet the stringent technical requirements and involve no problems of town and country planning, avoid the use of good agricultural land and are acceptable to the public at large (p. 653).

The emphasis on environmental aspects has greatly increased since the 1960s. Gammon (1981) identifies three major aspects: ecological, amenity, and socioeconomic. When the early stations were sited emphasis was mainly focused on those environmental effects that might influence the operational behaviour of the power stations.

10 A question of balance

Of these criteria, 5 are given particular emphasis: transmission grid, flat land, foundation conditions, water, and acceptable environmental effects (Gammon, 1982). It is important to appreciate that these siting criteria are generic considerations, that is to say that any short-listed site can comply with. However, the selection of short-listed sites is not determined solely by these considerations, but also reflects other factors that the Board may have in mind when making a decision.

It should also be noted that few of the criteria are absolute constraints, the exceptions being cooling water, population limits, and perhaps foundation conditions. Most are relative criteria that can be varied at a cost measured in economic or environmental terms. In selecting a feasible site the Board clearly has to reach a balance between their statutory duties to maintain an efficient, integrated, and economical electricity supply *and* the need to preserve the environment. This is a very difficult task. As Gammon and Pedgrift (1983) comment, 'The overall assessment of the technical, economic and environmental aspects between one site and another is not easy' (p. 45). Problems arise from the evaluation of different combinations of environmental effects at each alternative site. Some effects can be accurately predicted such as noise and thermal pollution, others require a wholly subjective judgement such as the visual effects of the plant and transmission lines, yet others depend on computer prediction such as environmental accumulation of radionuclides.

The eventual decision will probably state that location 'x' is on balance the most satisfactory location based on an assessment of the technical, economic, and environmental suitability of the potential sites. How this is achieved in practice is not documented but it appears to involve verbal debate among Board members based around recommendations brought forward by the siting engineers. It is claimed that in making their decision the Board attempts to interpret what reasonable men might feel to be the right balance between economy and environment.

It seems, however, that there are no detailed formalised objective or numerical evaluations made of the alternatives to aid this decision making process by ranking candidate sites. It also seems that attention is focused on only a very small number of sites and that either none, or at most one or two, are studied in any great detail prior to the public announcement declaring the Board's interest in a few sites. The 'balanced judgement' argument only appears late in the site selection process, whereas logically it might well have been regarded as having a role in the early site screening process. There have also been some suggestions that the 'balanced judgement' argument is an attempt to justify retrospectively the choice of a site made for other reasons – for example, related to engineering matters. The CEGB would probably reply by pointing out that they try to avoid sensitive environmental areas during the site search phase and this does amount to a balancing exercise of some sorts; but it is clearly not the same as using landscape evaluation techniques to guide the search process in the same way that transmission grid analyses may be used to identify candidate regions. The counter-argument is that perhaps the Board already give too much regard to environmental matters, and they may feel that environmental controls are too stringent and are being misused by those who oppose nuclear power. Gammon and Pedgrift (1983) note that:

> It is now possible to worry that some controls may be provided more for political reasons rather than on scientific grounds or cost/benefit considerations (p. 45).

This statement might well be considered grossly misleading if it were to imply that detailed cost-benefit analyses should be used to determine sites, since environmental aspects are rarely capable of monetary quantification. What they probably mean is that the best sites should be selected based on engineering considerations and this selection implicitly involves an intuitively informed cost-benefit analyses. However, such an evaluation need not be scientific, and there are no reasons why it cannot be questioned on political grounds although this is something that the CEGB have so far avoided by their use of national interest arguments. It is also quite understandable that engineers would want to keep their decisions relatively pure by emphasising their objectivity and their scientific rationale; things which most politicians do not understand and are reluctant to comment upon. Currently it seems doubtful whether this era of gross political neglect and public apathy will last much longer.

Once a site is selected and the design plans made, the Board view their environmental duties as involving careful architecture and landscape design. With respect to the Sizewell B application it is stated that

> The design and setting of the power station in the countryside is therefore of importance. The CEGB is very conscious of the need to ensure the best possible appearance of its proposals (CEGB, 1982, p. 191).

They seek to achieve this goal by preparing a landscape plan, by seeking advice from the Royal Fine Art Commission, and by

> . . . the use of material and colour to enhance the characteristic shapes of the main engineering elements of the PWR. The objective would be to complement the rectilinear shape of the existing station when seen in the setting of the extensive seascape of the Suffolk Heritage Coastline, and its hinterland including the Area of Outstanding Natural Beauty (p. 191).

Extensive tree planting and hedgerow, tree, and shrub screens will also be used to surround the site with woodland; the costs presumably demonstrate the CEGB's concern for the environment. It is perhaps easy to overlook the fact that this site at Sizewell is located on a Heritage Coastline in an Area of Outstanding Natural Beauty, that the reactor building will stand +75 metres above sea level, and that the supergrid pylons are about 45 metres high (CEGB, 1982, p. 126). Clearly the CEGB are trying to do everything possible to minimise the visual effects of building a nuclear station at a particular location but some things are quite impossible. The cumulative effect of one or two more PWR's at Sizewell is also something that is currently overlooked, as is the landscape impact of decommissioning the A station.

A critical review

In offering a critique of certain aspects of the CEGB's siting procedures considerable importance has been attached to first establishing as accurately as possible 'what they actually do'. That this could be done at all is a tribute to one member of the CEGB's planning department who was prepared to describe in

67

detail their activities and procedures. A very detailed study of other aspects is provided by Drapkin (1974). However, not all aspects of the CEGB decision making process are accessible and some gaps in knowledge do remain, although the Sizewell B public inquiry has been a tremendous source of detailed information about siting policies and nuclear safety of the kind not previously available or published in the UK.

In general it is clear that the CEGB's siting procedures are thorough and professionally applied. However doubts can be expressed as to whether the basic concepts are the correct ones for siting nuclear plant and whether the techniques being used are the most relevant and modern ones. Whilst the Board try to discharge their statutory duties in the best possible manner, their attitude often appears to be high-handed, their siting methodology poor, some of their data suspect, and their interpretation of their statutory duties may be questionable and perhaps should be subjected to closer political scrutiny. As Drapkin (1974) points out, 'Electricity proposals carry with them an unanswerable assumption: electricity developments are essential' (p. 252). Thus electricity developments are viewed as being different from most other planning proposals in which the possibility remains open of questioning the need for the development. The other major difference is that power station developments are viewed as a national good, one that is largely immune to local protest. However, if the issues are primarily nationalistic in character then they require some forum for debate. The planning process is wholly inappropriate because it can only really consider local aspects of a decision made on the basis of national arguments.

It may be observed that the Board regard the consents procedure as largely an exercise in bureaucracy. If they adopt a site they clearly expect to build a station there once it can be justified, irrespective of all opposition. This seemingly high-handed attitude may be viewed as a reflection of their confidence in the thoroughness of their case, plus a degree of self-righteousness which the national needs arguments may confer upon the whole process, as well as a great sense of urgency. An illustration of this confidence can be seen in the view that stations for which consents are still awaited are sufficient of a commitment for advance orders to be made for parts which have long lead times. Whilst the Sizewell B inquiry was about half-way to completion in deciding whether a PWR could be built at Sizewell and whether it would satisfy the safety requirements, the CEGB have ordered about £100 million worth of hardware. Their hope is that it can be shown to be safe retrospectively (*Sunday Times*,

November 13th, 1983, p. 53). A similar slightly blasé approach to planning consents was evident in the early 1970s when blanket approval for multiple reactors was obtained without any explicit specification of reactor designs. No doubt similar thinking is involved in various governmental suggestions that the Sizewell B inquiry should be a generic inquiry for all subsequent PWR applications. It can also be seen in previous inquiries where the scope has been deliberately restricted to local planning matters; one imagines this will become fashionable again in the post-Sizewell B situation. To some extent this can be justified on the grounds that the existing planning system does not cover either the needs or the safety aspects of major developments. Planners will usually inform their members that these questions exist, but then pass all responsibility for them to the appropriate government ministry or to the developer.

The effect of the statutory planning process is mainly that of creating delays. At a time when there are no electricity shortages and no short-term forecasts of shortages, one would imagine that all concerned could take a more relaxed attitude to such matters. An outsider may well be excused for considering the CEGB impatience over Sizewell B must reflect other, undisclosed, factors; it has certainly nothing to do with the need for additional capacity. In the long run these pressures for swift decisions will result in a streamlining of the entire process and that will mainly serve to further internalise the important decisions to secret Board meetings. Local arguments have, perhaps sadly, no real hope of winning. To paraphrase a senior CEGB planning engineer, 'if we want site x and we can make a good case out for it then nothing other than technical factors or a change in the national economic situation will prevent us from developing it, if we want to'. There is nothing intrinsically wrong with such a view but it does serve to emphasise, yet again, that the site selection process has to be seen in the wider national context. This implies that it is the initial site search phase that requires more explicit public scrutiny. At present this process is hidden from view and largely exempt from either political or public participation. The decisions made here are the really fundamental ones from a geographical point of view; everything else that follows is no more than satisfying various bureaucratic procedures given a particular site.

Gammon and Pedgrift (1983) note that

> With the passage of time much of the country has been considered for sites and the CEGB now hold a pool of prospective sites, some partially developed, from which to select the most suitable (p. 41).

The fundamental question concerns not the broad criteria used for this selection process but the nature of the evaluations that are applied when comparing sites and the weightings given to the different criteria. The suspicion is expressed that engineering rather than good planning practices dominate the selection process. Drapkin (1974) suggests that perhaps a desirability coefficient could be computed for each potential site that incorporates the competing interests of safety, environment, and economics of location. The ideal planning process, according to him, would be to rank sites to determine the order of priority. He considers that 'The current, humanly inefficient process, approximates this end' (p. 253). On the other hand any such rankings would at best be partial, based on an incomplete set of potential sites, and would at present be done without recourse to any formal evaluation technique. It will be demonstrated in a subsequent chapter that the supply of potential sites is not as restricted as the CEGB would have us believe. The current habit of short-listing two or three sites, usually very close to each other, prior to the adoption decision is very reminiscent of planning practices in the 1950s; it is certainly not what modern planning decisions should be based upon. Nevertheless, it does have some very definite advantages: the 'lack of alternatives' argument provides a powerful reason for selecting the preferred site; it avoids political debate, it avoids the need for expensive detailed comparative studies of alternatives, it precludes the use of explicit optimisation approaches in making the siting decision with mathematical programming methods being used to handle the trade-offs between the conflicting criteria, and it keeps the entire exercise simple and readily explainable to the public. Yet no professional planner or geographer could possibly commend it for use in the 1980s. There is a real risk that sites will be selected for reasons which are inherently irrelevant to the uses for which they will eventually be put.

The geographical view is even simpler. If the full range of *all* potential sites are not explicitly considered, then the national interest might be being dealt a great injustice. It precludes the ranking of sites on a national basis. It prevents any rational discussion about the prospect for isolated sites and it ignores the spatial or locational freedom that is available for siting nuclear power plant even in the UK. If the CEGB use inefficient spatial search methods here then everything else that follows may be either wrong or inefficient when judged in the national context. This is not to say that comprehensive site searches are not performed but that to be seen to be done properly in the 1980s it is necessary to build up the computer data bases necessary for detailed multiple criteria searches. There is no evidence that the CEGB have such data, or are indeed interested in the application of such methods. This neglect is fundamental if there were to be, as indeed there should be, a greater need for site optimisation with respect to safety criteria.

Various aspects of the current siting process require further detailed comment in relation to current practices.

Best site, feasible site, or convenient site?

Some of these problems can be crystallised in the question as to whether a particular location is the best site or merely a feasible site or simply a convenient site. Openshaw (1982b) argued that convenience rather than public safety has dominated the choice of sites for nuclear power stations in Britain. This evoked the following response from the CEGB's Director of Corporate Strategy. In a letter to *The Times* (4 August, 1982) he wrote:

> It really is nonsense to say that convenience rather than public safety has dominated the choice of sites for nuclear power stations in Britain. Those of us involved in the search for sites and responsible for making recommendations to the board about their selection have to make a balanced assessment of all relevant technical, economic, environmental and safety considerations because these elements form part of the statutory duties upon the board in the conduct of its business. Each site so selected is subjected to public scrutiny through the necessary consents procedure. . . . Whether or not a site is acceptable from the safety viewpoint is determined by the Nuclear Installations Inspectorate of the Health and Safety Executive – not the CEGB (C.A. Davis, p. 11).

A distinction should be made between the sites that satisfy the existing safety requirements (see Chapters 3 and 4 in this book), and sites which explicitly seek to maximise public safety. Very few of the existing nuclear power station sites in the UK would satisfy this latter requirement. The matter of 'public scrutiny' has already been dealt with. The point here is that once the Board applies for a planning consent the decision is either 'Yes' or 'No'. If the proposal is rejected there is no assurance that the rejection is the best solution. Nor is unsuccessful opposition an indication that the Board's solution is the best.

The argument that a station is needed, and the choice of a particular site need to be separated. If the demand argument for more generation capacity is accepted and the choice of a nuclear plant is made, then the subsequent siting decision should be based on a study of alternative sites in the national context. This study of alternatives should be an explicit part of the proposal so that the Minister, aided by a public inquiry if need be, can make the final locational decision. This would be far fairer than the present system in which the national needs argument is conveniently linked to the choice of a particular site.

The need to open up the locational decision and the evaluation of alternatives is important because identifying the best site or the most desirable site is far more difficult than demonstrating a national need for either more power or cheaper power generation. This is an issue of considerable practical significance and it is something which should not be left with the Board to determine.

Public inquiries

It is precisely this point that Mr J. Popham of the Suffolk Preservation Society made in his cross-examination of Mr J.W. Baker, secretary of the CEGB, and Board member, at the Sizewell B Public Inquiry in 1983. Popham was in the fairly unique position of having been allowed access to the Board's internal development reviews and associated papers relating to siting. His questioning of Mr Baker is from a position of considerable knowledge about Board internal affairs. The key question he asked is:

> Mr Baker, is it the Board's case that Sizewell is merely a suitable site for the first PWR, or is it the Board's case that Sizewell is the best site for the first PWR? (Sizewell B Public Inquiry, 1983, p. 48).

Baker's answer was:

> It is of course, the Board's case that it is a suitable site. Of those which we considered, it was possible to demonstrate that there were indeed other suitable sites, but *taking a rounded view of matters*, Sizewell was our preferred site, *taking a balance of the judgement of the factors involved* (Sizewell B, 1983, p. 48).

The critical statements have been placed in italics. A reference was then made by Baker to the CEGB's proof of evidence which, it is claimed, states that 'Sizewell represents on balance the most satisfactory location'. Popham then tried to extract from Mr Baker evidence to justify this statement. Eventually Baker admitted:

> . . . you may well be right that none of the witnesses comes to

a specific sentence in which he says that Sizewell is the best site in national terms (Sizewell B, 1983, p. 51).

Popham then tries to identify the reasons as to why Sizewell was chosen as a site for the first PWR. He establishes that the Board first decided to select Sizewell in 1978, a decision later confirmed in 1980. It was only after this date that evidence was produced to justify the decision. He quotes from a CEGB internal memorandum date 1981 – nearly 3 years after Sizewell was first selected and almost a year after the decision had been confirmed – which talked about the need to justify the decision at a public inquiry in the face of close examination. Further details of the Board's post-decision-making justification emerged later.

Apparently the Board took 3 factors into consideration in selecting its short list of possible PWR sites: (1) there was a safety requirement of the Nuclear Installations Inspectorate that a 'remote' site should be used because this was government policy for the first reactor of a type new to the UK. (2) The desirability of minimising pre-construction work and permitting a quick start. It was claimed that an early start at Sizewell would save £48 million because of previous preparatory work at that site; this allowed a 6-month time advantage over other sites. (3) The desirability of reducing objections, so that there was a focus on existing sites where the local people already accepted nuclear power (pp. 54-5, op. cit.). This latter point was thought to be a major consideration in the 1978 decision to select Sizewell. In addition there was no need for extra transmission works or new supergrid lines.

Popham then tried to discover the nature of the evaluations that led to the original short list of sites. He asked Baker a number of questions.

Popham:

. . . did the Board compare the capital cost of developing the alternative sites when selecting the shortlist? (Sizewell B, 1983, p. 58).

Baker's answer:

. . . no. . . . Differentiating by site in capital terms was not a consideration (pp. 58-9).

Popham:

. . . did the Board compare the net effective cost of the alternative sites?

Baker's reply:

. . . no . . . (p. 59).

Popham:

Did the Board make a full site comparison of the two benefits and system operational factors when selecting the short list and considering the sites? (p. 59).

Baker:

. . . no . . . although there was a general awareness of the power flow problems.

Popham:

. . . did the Board consider the quality of the landscape of the alternative sites? (p. 60).

Baker:

. . . no.

Popham:

. . . did the Board consider the visual impact of the proposed stations on their respective sites? (p. 60).

Baker:

No.

Popham:

Did the Board consider the effect which the proposed stations might have on the ecology of the area? (p. 60).

Baker:

Not specifically.

Popham:

Did the Board pay regard to any official designations such as areas of outstanding natural beauty, heritage coasts, and the like when selecting the sites? (p. 60).

Baker:

No.

Popham then draws attention to Section 37 of the 1957 Electricity Act and notes that in their proof of evidence the CEGB omits part of the first sentence. The missing phrase is in italics. The act states 'In formulating *or considering* any proposals . . .' (CEGB, 1982, p. 4). Popham argues that the formulation process must include the original decision, but there seems to have been no detailed comparison of the merits of alternative sites, or of landscape quality, or of visual impact, or of ecology, or of official designations during the consideration stage, and indeed in this instance in the formulation stage as well. He then suggests that technically the Board must have been in breach of its statutory duty when it made the decision to select Sizewell without any evaluation of these factors.

Baker's response is very interesting because of what it reveals about Board decision-making. In the cross-examination he could not simply avoid negative responses to Popham's questions because Popham already knew what the answers would be from his reading of internal Board documents. Baker said 'The original paper in 1978' that led to the first decision to select Sizewell

> . . . is essentially a technically based analysis of the suitability of the site against a power station at that stage no more than, if I can put it this way, a concept on a piece of paper. There was not an organised design. There was not detailed engineering. The basis of the papers was to examine sites for a design which would need to be developed.

The next section of the reply is even more interesting and revealing about how the Board interpret its statutory duties with respect to the environment.

> . . . the further and more important point is that those in the Board whose job it is to examine sites and to bring them to a stage where they can be considered, also are the people who, in relation to the construction of new power stations have the primary interest in ensuring that as the site develops we carry out our duties under Section 37. What I am suggesting, therefore, is that their familiarity with the proceedings would mean that they would be conscious of these items in formulating proposals (p. 62).

So it seems, quite incredibly, that a site is first adopted and then detailed investigations are carried out. The Section 37 responsibilities are mainly relegated to the project development stage. Indeed Baker states

> . . . the Board is fully conscious of its section 37 responsibilities, some of which can be exercised at a desk stage and some of which obviously can only be exercised as the project develops, and that the people who carry responsibility for investigating sites, putting forward propositions, are fully aware of their responsibilities in that area (p. 70).

The question as to whether this is sufficient and to what precisely these responsibilities are in practice really requires answering in some detail. Their approach seems to leave much to be desired, in that in the ideal world the Section 37 responsibilities would mainly influence the site evaluation process that leads to the selection of a site, and not a duty that largely begins once a site has been adopted. This is, in fact, a statutory requirement if the 'formulating' and 'considering' phrases in Section 37 are regarded as covering the site evaluation and selection processes. It can also be argued that the informal way in which these duties are discharged is hardly a satisfactory approach. Indeed it seems that in the case of Sizewell, a broad familiarity with the various environmental and ecological backgrounds was taken for granted even though as Baker admits '. . . the paperwork does not cover the particular items' (p. 63).

It would appear that a subsequent examination of alternative sites in south-west England did in fact include some environmental details, to the extent that there was a printed summary of environmental factors against each of the alternative sites. This it seems, is amazing, regarded as sufficient to fulfil the Board's statutory duty with respect to environmental matters. It could be argued that this may satisfy a strict interpretation of Section 37 but it can hardly be regarded as either adequate or satisfactory. Nor is it clear how sites can be selected or short-listed without detailed environmental surveys.

What appears to happen is that the Board is presented with a siting package prepared by the CEGB's planning department and that the executive decision is greatly simplified by the prior site selection done by the planning department. Throughout, and at all levels of decision-making, there seems to be a high degree of reliance on professional judgements, opinions, impressions, and other fairly soft qualitative information. There seems to be a virtual absence of detailed comparisons using quantitative or qualitative measurements of the sort that one would expect to be used in evaluating the suitability of sites for major capital investments.

According to the evidence unearthed by Popham it would seem that matters of technical convenience predominate over all others in the selection of sites. This may be justified on the grounds of pragmatism and a previous lack of critical interest by outside bodies in the siting process. However, Sizewell may be unique. It is unique in that the public inquiry has exposed many aspects of a hitherto unpublicised process, and it may also be unique in that the conclusions drawn from it may not be relevant to other siting decisions. Moreover, it is hard not to have some sympathy with the CEGB. Their siting procedures have evolved over a long period of time during which period they have been subjected to very little if any critical comment. They do far more than might be expected from other large industrial enterprises and they can argue with some justification that the problems they face concern broader issues about how national interest siting decisions should be made. At present it is their task to do this and clearly they do the best they can to grapple with the difficulties and provide considered solutions. Any major changes to their procedures would probably require an explicit governmental directive and the absence of any such concern may be taken to indicate that they are doing a good job. Others may well hold a less complimentary view.

Handling of environmental matters

It is important to note that the CEGB (and the SSEB) are not required to develop *only* the cheapest possible system nor need they avoid all damage to the countryside. Instead they are faced with the task of establishing the most proper compromises, or as it was put for Sizewell 'a decision in the round'. It is claimed that every effort is made to avoid protected areas of landscape and amenity value although it is not always possible. The Countryside Commission (1968) estimated that some 60 per cent of the England and Wales coastline is given statutory protection as national parks, green belts, areas of outstanding natural beauty, nature reserves, etc. An additional 10 per cent is already fully urban in character and only the remaining 30 per cent is available for development. it is, as Hunt (1970) writes,

. . . a tribute to the care that has been taken to avoid encroaching on these areas that only two of the power stations developed in the period under review (1961-78) have had to be sited wholly within protected areas (p. 654).

However, this overlooks the fact that many of the remainder are located close to protected areas or are in areas prized for amenity. It is very easy to underestimate the strength of concern that the Board evidently shows for avoiding technically suitable sites in the most beautiful areas. At the same time one can ask important questions about how the Board make their landscape evaluations.

Gammon (1981) writes

As far as possible the Board avoids technically promising sites in the most beautiful areas such as the Northumberland coast

77

area of outstanding natural beauty (AONB). But this does imply that the Board's interest in the use of areas which have somewhat lower quality, such as Druridge to the south outside the AONB, should be understood (p. 250).

It has been stated in public by a member of the CEGB's corporate strategy department that a very simple procedure resulted in the selection of Druridge Bay. Once it was decided that a nuclear power station was needed in the Northern Region of the CEGB a search was made using aerial photographs of the coastline from Scarborough north to Berwick. A small number of possible sites were identified but of these the first site that satisfied the various criteria was Druridge. A counter-view might be that Druridge was selected because it happened to be conveniently located near two spurs of the supergrid and allowed the possibility of multiple developments. Its location in an area where industrial developments would not normally be permitted meant that it would easily satisfy the existing siting criteria of the Nuclear Installations Inspectorate. The question arises however as to why Druridge should have been selected at all. Any development there would conflict with the existing statutory plans for the area, and whilst it is not protected for its landscape quality it is certainly prized as a recreation and amenity area for the neighbouring urban areas of Tyneside. The implied *threat* in Gammon's statement is that should Druridge be denied then the CEGB would move up the coast into areas of outstanding natural beauty in the search for alternative sites. In fact the best alternative sites are probably considerably further to the north, they would be more expensive to develop requiring new roads and new transmission facilities; there would be problems with labour supply, housing and staff recruitment, and with the provision of necessary local services; and strong opposition might be expected from environmental groups. In this instance their desire to avoid the AONB may be a reflection of convenience and have little or nothing to do with landscape evaluation.

Another example is in south-west England. One possible site for a nuclear station is on the river Fleet at Herbury, inside the Dorset AONB just inland of Chesil beach. It could be direct cooled and the buildings could be well screened by digging the station into a small valley; see Gammon (1981) p. 250. The other site is inland at Winfrith. This is outside the AONB but is so located that the cooling towers and their plumes of vapour would be visible over a wide area; nevertheless, it would leave the coast unspoiled. The implication here is that an AONB site would be 'better' than the alternative. The CEGB investigated four other sites in the south-west (CEGB, 1980); at Innswork Point, Bugle-Luxulyan, Gwithian, and Nancekuke. In 1982 it was announced that the Gwithian and Nancekuke sites could not be considered for future nuclear power stations because of unsatisfactory geological conditions (CEGB, 1982); they are also, incidentally more than 30 km from the supergrid and it may be suspected that for this reason these sites were never seriously considered. The site at Innswork Point is being reserved for a possible new coal fired station mainly because there are too many people too near to the site to satisfy existing population siting criteria for nuclear plant. However, this is a public relations ploy because there is no real prospect of a coal or oil fired plant being built there. The site at Winfrith Heath in an area of a highly interesting, perhaps unique, ecology was relocated to within the existing UK Atomic Energy Authority site located there. The intention is to proceed with the Winfrith AEA site with Luxulyan and Herbury held in reserve. The CEGB statement explains the decision like this:

> The site considered at Winfrith Heath in Dorset, adjacent to and to the west of the existing UK Atomic Energy Authority establishment and including land owned by the Authority, would be technically suitable for power station development. However, the Board considers that an area largely within the site of the UKAEA establishment itself would be more suitable for power station construction purposes, taking into account ecological aspects (CEGB, 1982, p. 1).

In some ways this decision is also very convenient. The Winfrith UK AEA site already has grid connections and the close association with the AEA suggests that it could well be the eventual location for the first commercial fast breeder reactor to be built in the UK. Of course there has been no public announcement to this effect and the CEGB currently have no such plans (by which they mean any commitment to develop a fast breeder reactor). Nevertheless it should be noted that Winfrith was in fact the preferred site for the Dounreay prototype fast breeder reactor. It was considered sufficiently close to London for prospective foreign buyers to be able to easily inspect it. The site at Herbury, so favoured by Gammon (1981), turns out to be suitable for only one reactor if it were to be shielded to minimise visual intrusion; two or more reactors would be considerably more prominent. The development at Luxulyan could be hidden to some extent by the china clay workings, but it would still be dominant over a large area and there could well be difficulties in obtaining adequate cooling water supplies. Finally, it is also seen as constraining

. . . the pattern of additional generation within the South-West Peninsula in the longer term if major overhead line reinforcements both to and within the peninsula were to be avoided (CEGB, 1982).

The implication is that developments at Winfrith would be more convenient than the AONB sites. So is this yet another good example of the Board's concern for protecting the environment, or merely a rationalisation of a site adoption made now but intended for a currently non-existent reactor type? Furthermore, the apparent concern for avoiding the need for line reinforcement or duplication of the supergrid may only be a short term feature; multiple developments at virtually any of these sites would eventually require it.

Further information on the Board's handling of environmental matters was unearthed during the Sizewell Inquiry. On Day 247, Gammon of the CEGB describes what happens:

> When a site is adopted, it is regarded in the Board's collective viewpoint as being technically suitable and that developments on that site would have environmentally acceptable consequences. Now that view is tested when we apply for consent, especially in situations where there is a Public Inquiry (Sizewell B, 1984, Day 247).

It is the Board's decision, but their method of arriving at a balanced decision is somewhat suspect. Sizewell is an environmentally acceptable site in 1984 because it was considered acceptable in 1957. Yet the site lies within an Area of Outstanding Natural Beauty, it is partly within a Heritage Coast area, there is an RSPB reserve to the north, a Nature Conservancy Council's National Reserve 5 km to the north, Sites of Special Scientific Interest 3 km to the south, and a new proposed SSSI includes part of the site. The CEGB, however, still regard it as being environmentally acceptable because it is not in a National Park nor on National Trust Land, nor common land, nor in a green belt, nor of high grade agricultural land (Gammon, Sizewell B 1984, p. 17, Day 248). On the other hand, the principal environmental justification for Druridge Bay is that it does not lie within an Area of Outstanding Natural Beauty. Clearly there has to be some kind of trade-off during the site evaluation stage and a detailed evaluation made of alternative sites and alternative locations. It is probably apparent that the principal reason why Sizewell is still regarded as environmentally acceptable is because of the existing nuclear power station development there; and that is a very weak excuse, especially as there may well be more satisfactory locations from an environmental point of view which can offer similar or better system benefits compared with Sizewell. The problem is that there is at present no formal set of criteria which could be used to quantify current concerns and thus evaluate alternative locations in a rigorous manner.

Attitudes towards public participation

It would appear that, in practice, environmental arguments may be being used to reinforce the use of locations that are preferable on economic and engineering grounds and as an excuse for not adopting a broader and comprehensive site evaluation process. The key decisions are made by the CEGB meeting in private. There are opportunities for the public to express their opinions, for local councillors to state their views, and consultations are made with large numbers of government agencies; there may even be a public inquiry; but at the end of the day the CEGB expects that its siting decision will be approved. One might well be excused for considering that they have a very narrow and peculiar view of planning and public participation. They clearly see no need for public participation in the fundamental siting decisions, and take the view that technical objections are the only ones that matter. People who disagree with the decisions are either regarded as being misled or advocates of a return to a peasant subsistence economy. The impression is given that the Board is *right*, that their decisions are *right*, and the only real task is to explain why they are *right* to the local communities affected by their developments.

Gammon and Pedgrift (1983) note that the Board now anticipates both active and passive opposition to development proposals. Their response has been to seek '. . . more effective communication with local people' (p. 43). The problem is neatly summarised by Matthews (1977) when he writes:

> Despite all the technical literature and information available, there is serious difficulty in communicating with the general public on highly scientific matters in a way which will gain their interest and trust, and enable the problems to be seen in a proper perspective against the background of everyday risks and hazards (Matthews, 1977, p. 8).

It would seem that to date this propaganda campaign has not succeeded, and there are indications that public acceptability of nuclear power is diminishing as the public become more critical about CEGB statements. The CEGB, of course, see things quite differently. Mathews writes:

> The CEGB's policy of frankness has paid dividends, for although nuclear opposition groups have tried to stir up local opposition at nuclear stations their efforts have met with little success (p. 5).

This may be due to the absence of any channels for effective complaints. It would appear that most civilian neighbours of nuclear power plants are worried but prefer to believe the assurances given by the CEGB that their safety is guaranteed. Indeed one cannot help but be impressed by the absolute faith that CEGB employees express in public meetings that nothing untoward will ever happen at their nuclear station. Whether this is a conditioned response or a reflection of professionalism of the highest order, or both, is not readily determinable. The anxieties of local residents may well be greatly reduced by such performances, but it may be a fragile faith that is constantly being disturbed by the continuous media coverage of all nuclear incidents no matter how small or irrelevant.

Whilst this may be reassuring, this is not really what public participation is all about. The only effective inputs can be during the site selection and evaluation stages. These inputs will only have an effect if the CEGB were to have a statutory duty to take them into account when formulating or evaluating decisions. This requirement exists for all major town and country planning exercises in policy formulation: it should also be applied to power station planning.

Engineering priorities come first

It is apparent that the CEGB's interpretation of their statutory duties gives maximum priority to engineering matters and to an engineering perspective when making balanced judgements. Consider for example the general practice of developing existing sites where possible and practicable and if they satisfy the electrical supply requirements; but they are the sole arbiters of all three decisions. Gammon (1982) states:

> By developing such sites to their full capacity, as determined by environmental and technical limitations, advantage can be taken of existing facilities – such as transmission outlets and improved local roads – minimising the amount of new works required. Another benefit is that it reduces the overall number of greenfield sites required for power station development (CEGB, 1982, p. 3).

But what precisely are these 'environmental and technical limitations'? By the former is meant factors such as the capacity of the local heat sink to absorb more waste heat. It does not mean that visual or landscape considerations will ultimately stop development. The technical limitations involve such things as the capacity of the existing transmission lines, adequate site space and site conditions, and possible problems of simultaneous developments on the same site. But it should also be noted that the redevelopment and full development of existing sites is virtually guaranteed by the various Nuclear Installation Acts. This arises because the site operator has a statutory duty to restore any obsolete site to its previous condition with only a normal background level of radiation. This is probably impossible with present technology and it could well involve a timescale of several hundred years. It is far easier to postpone for as long as possible the major problems of decommissioning old plant, mainly to ensure that these problems do not coincide with the public debate about the safety and desirability of nuclear power. The problem can then be effectively hidden by redeveloping the site. The old reactor hulks can simply be left encased in concrete, probably painted green for environmental reasons, and left until such time as complete dis-assembly becomes either feasible, or economic, or necessary.

It is true that the number of new greenfield sites may well be reduced but the final number of new sites required depends on the eventual size of the nuclear power programme, and this is a highly uncertain quantity. Indeed the corollary of both arguments is that new sites would have to be large enough to both house multiple developments, and make possible the emergence of major reactor parks with the collocation of thermal and breeder reactors with reprocessing and fuel fabrication facilities. It is not known how many of the existing sites could be extended to accommodate such developments, but on the evidence of Chapter 1, such a need may well emerge soon after the start of a fast breeder programme; probably the early years of the next century. So it seems that sites originally selected for small thermal reactors may come to house far larger multiple developments of the kind not clearly envisaged when the original site evaluations were performed.

Incremental decision making in the direction of an undisclosed objective

It can be argued that the long time horizon of the site development process has allowed the CEGB to adopt an

83

incrementalist approach to planning. The application for a particular site may be for one reactor and the justification for its selection may focus on the role of that reactor in meeting certain national energy needs. The subsequent B, C, and D etc. stations follow after a time interval because of the desirability of fully developing the site. However, it may be that the site would never have been selected in the first place if the full development process had been either known or anticipated. It is, of course, far less likely that plans for the subsequent stations would be rejected once the initial station had been approved. This is one example of the way in which the planning system can be manipulated by an incremental decision-making process over an extended period of time.

The solution here must be to consider not merely the terms of reference of the initial proposal, but also the possible fully developed state of the site. The CEGB would probably point out that they have no such plans for full development or if they do they do not know either when or with what reactor types. Prudence would require them to remain flexible about such matters. Nevertheless, the prospects for multiple developments will certainly have been considered during the site evaluation process. Only a little additional imagination is required to suggest that some of the principal criteria may themselves have been engineered over a period of time. For example, the routing of the supergrid has itself increased the prospects for certain sites being developed. The question arises as to whether or not its location was deliberately influenced by the prospective future development potential of certain areas. It would certainly be sensible to make such arrangements even if they do prejudge subsequent siting decisions. There is admittedly no evidence for this but it may attribute a degree of optimality in Board decision making that may simply not exist. Alternatively it may be that they have been successful in operationalising long-term strategic plans, and they cannot be criticised for that. What is needed, however, is a different planning procedure which can fairly handle major long-term developments of this sort.

A softly-softly incremental approach may well be the optimal strategy for the Board to follow; indeed it may be the only one that fits in with government fuel policy which mainly influences the order in which stations can be built, rather than their actual location. Indeed the only real open-ended part of the planning process is the chronology decision about when to develop a particular site. This arises because the Board has been very successful at giving the impression that there are very few suitable sites. This incremental approach also reflects the absence of any clear national policy for the development of energy resources, including the role of nuclear electricity. Although there is currently government support for nuclear power it has not ruled out coal fired plant. Instead of a long term plan or policy the Board is left more or less to do whatever it believes is necessary to fulfil its statutory functions. Each new proposal is considered on its own merits. Ideally it might be expected that this development process would fit into a policy framework approved by Parliament. The generating boards might well have been given guidelines for achieving particular targets – for example, a 60:40 balance between coal and nuclear by 2010 – or told what balance might be considered desirable. The guidelines might well be reviewed at 5-year intervals, or as circumstances change.

The Board describes in its Sizewell B evidence a high nuclear option for the year 2010, with 70 per cent of nuclear electricity, although there is no way of knowing whether this level of fuel diversification is desirable or even attainable; in 1981-2 coal accounted for 84 per cent and nuclear for 11 per cent. Yet no new coal stations have been built for over 10 years. Unless new coal stations are ordered soon then the proportion of coal fired electricity generation is estimated to drop to less than 10 per cent after 2010. So although there are no formal CEGB or Parliament-approved plans for an all-nuclear future, this is, in fact, the implicit objective of current Board planning. Their declared pro-nuclear stance – for example, Mr Baker declared in 1981 'I'd like to make it clear, if it isn't already, that the CEGB is pro-nuclear' (p. 1) – only serves to emphasise their intentions, although of course Sizewell B apart there is technically no commitment to anything. Whether this is a sensible way for any government to delegate its corporate planning powers in such an important area of public concern, is an interesting question. If the intentions were deliberately kept secret to avoid public and political debate they could be strongly condemned. When one cannot be certain that there are any long term intentions then this is also very undesirable. If no one on the Board has a clear picture of the future that the incremental decision-making is meant to produce, then they should all be sacked! Yet if they do have such an image it should be open for public discussion and debate in sufficient time for society to judge whether it shares the same goals as the CEGB. Instead it would seem that both public and politicians are not to be trusted in case they make the 'wrong' decision.

Conclusions

The CEGB operates an extensive planning process which is applied in a thorough and professional manner. The only real criticisms are as follows: the process has virtually no opportunity for effective public participation; it does not allow for the detailed evaluation and comparison of alternative sites; it is based on informal evaluations; it seemingly gives preference to engineering aspects with environmental considerations mainly relegated to the minor details of site development; it is based on an incrementalist approach to power station planning in which the long-term objectives, if any exist, have not been publicly defined; and there are real doubts as to whether the full range of possible candidate sites have ever been identified or considered during site searches. The national interest arguments that have been extensively used to justify site selection and power developments have allowed no opportunity for meaningful political and public debate of the basic assumptions, yet the emphasis on the national interest means that some method is needed of identifying those sites which are really in the best national interest.

Prudence dictates that public safety should be a major criterion in site selection. This is currently readily dismissed as involving no more than conformity with the Nuclear Installation Inspectorates population-siting criteria. Sites that satisfy these population restrictions are in the CEGB's view safe. There is no need to look for safer sites because their nuclear stations are safe. Others may well agree but still consider it sensible to look for even safer sites within the engineering restrictions that are currently applied. The view here is that the CEGB and the SSEB are neglecting their national responsibilities by failing to use siting as an additional, active, safety measure. Perhaps if they no longer had limited liability in the event of accidents, and if the Board members were regarded as being personally liable for civil damages, they might well adopt a less complacent view.

3　Remote siting policies in the UK

Before examining the selection of sites for civilian power reactors it is useful to examine the locational practices used for the preceding military reactors. It is worth emphasising that both the British and the US involvement in nuclear energy were initially for purely military reasons. The British 'atoms for peace' programme was in reality a crash programme to develop atomic weapons, even if some of those involved had dreams about more peaceful applications. A basic chronology of events is given in UKAEA (1979) with excellent historical accounts in Gowing (1964, 1974) and Pocock (1977). The historical details are interesting because of the insights they give as to why particular sites were selected, the criteria that were used, and the steps taken to ensure public safety.

Military reactor siting

The first nuclear installations were typically located in great haste under conditions of immense secrecy to meet urgent military objectives. The sites were selected at a time when knowledge of nuclear power was incomplete and experience minimal. Yet these locational decisions made over 40 years ago, often in a situation of great ignorance, are still important and the sites so selected still house major nuclear installations. Perhaps miraculously, post hoc and retrospective evaluations seem to have generally validated these siting decisions.

In Britain a natural uranium reactor was considered to be the only practicable means of producing plutonium for atomic bombs. It had been intended to build similar water cooled reactors to those at Hanford in the US, but this plan was altered when safety studies were made of the proposed reactors. Only one suitable isolated site was considered to exist at Morar in

Invernessshire. It was decided instead to use air cooling.

A number of key locational decisions were made in 1946-7. The UK Atomic Energy Act of 1946 created the Atomic Energy Authority and gave a tremendous boost to the drive to create a British atomic weapon. The US Atomic Energy Act of 1946 had effectively deprived Britain of US bomb secrets. Harwell (a former airfield near Oxford) was selected to house the reactor research group under Cockcroft and construction started on two reactors BEPO and GLEEP. A former munitions factory at Risley became the location of the atomic production group for the design and engineering of nuclear weapons, under Hinton, who was later to become the first chairman of the CEGB. A uranium metal plant was built on the site of a disused poison gas factory at Springfields, and two simple, air-cooled plutonium piles were built on the site of a former TNT factory at Seascale (Windscale) in Cumberland – the site was announced in July 1947. Apparently the Windscale Piles were first planned to be built at the Royal Ordnance Factory at Drigg but the location was moved to Sellafield when the site became available. Seascale was a particularly convenient location, which also housed a chemical separation plant needed to recover the plutonium. This required a coastal location for the easy disposal of liquid wastes, while the Drigg site became a dump of low level wastes. The Windscale Piles went critical in 1950-1 and the name Seascale replaced by Windscale to avoid confusion with Springfields. The chemical separation plant commenced in 1952.

A gaseous diffusion plant for uranium enrichment was built on an old ordnance factory at Capenhurst; the site was announced in 1949 and the plant commenced operations in 1953. Finally, in 1950 the Aldermaston site was taken over for weapons work.

As early as 1947 and certainly by 1952 there was a design for a reactor that would produce both electricity and plutonium. It involved a 40MW(e) carbon dioxide cooled graphite moderated natural uranium reactor known as PIPPA (pressurised pile for producing power and plutonium). It should be noted that the waste heat from the plutonium piles at Windscale was vented to the atmosphere after passing through a series of filters known as Cockcroft's folly, since they were not considered necessary; it was these filters that trapped most of the radionuclides released during the Windscale fire of 1957. Anyway it was thought that greater plutonium productivity could be obtained if a more effective cooling system could be developed. The demand for plutonium doubled in the early 1950s and the PIPPA design was optimised for plutonium production with electricity as a by-product (Gowing, 1974).

In January 1953 a government white paper announced what it termed 'A programme for nuclear power' (Cmnd 8986). There was to be a full-scale nuclear power station based on the PIPPA design, and an intention to develop a fast breeder reactor as an alternative route to acquiring plutonium. It also made sense in that the known resources of uranium were comparatively small and it was felt, even at this early date, that it was pointless aiming at a large-scale commercial power programme before it was known that uranium could be used economically, and this required a balance of fast and thermal reactors. The site for the first nuclear power station was adjacent to the Windscale site; the only alternative location mentioned at the time was East Yelland. Work commenced on Calder Hall A station in 1953 and the station was opened in 1956 (Kay, 1956). The site for the first fast breeder reactor, a 14MW(e) device, was also selected at this time. The chosen location was at Dounreay on an airfield near Thurso in Northern Scotland. The site was announced in 1954 and the reactor went critical in 1959.

The Atomic Energy Act of 1954 gave all responsibility for nuclear energy developments to the UKAEA. In June 1955 two additional reactors were announced for the Windscale site and a second station of four reactors at Chapelcross in Dumfriesshire, Scotland. The design of the Chapelcross reactors was basically the same as that for Calder Hall but the site was better and there was no need for raft foundations.

Not until February 1955 was there any public announcement about plans for a civilian nuclear power programme when a White Paper, 'A Programme of Nuclear Power', was presented to Parliament (Cmnd 9389). The power stations were to be run by the electricity authorities but the Ministry of Defence insisted on design modifications in 1958 so that military plutonium could also be produced.

Subsequent UKAEA developments are based on their existing sites with two new sites being announced. A prototype Advanced Gas-cooled Reactor (AGR) of 33MW(e) was announced for Windscale in 1958; it operated in 1963. A new site at Winfrith Heath was announced in 1957. This is the location for a high temperature gas cooled reactor (DRAGON), there being no room at Harwell for it. Full power of 20MW(t) was reached in 1966. In 1963 there was an announcement of a 100MW(e) prototype Steam Generating Heavy Water Reactor (SGHWR) also at Winfrith, full power being obtained in 1968. In 1966 a prototype 250MW(e) fast reactor (PFR) was announced for the Dounreay site with associated fuel production facilities being built at Windscale. Full power was not achieved until 1977, the year in which the first Dounreay reactor was finally shut down. Another new site was the selection of Culham in 1960 as the centre of thermonuclear research. Figure 3.1 shows the location of these early nuclear establishments.

These early sites were selected in secret. The true purpose was seldom announced, and there was little or no formal negotiation with interested bodies. The UKAEA are exempt from planning controls. At the time when most of these sites were chosen there may well have been no realistic conception of the possible public safety hazards that reactor accidents might have caused. The sites were selected in seemingly great haste and without any in-depth environmental surveys. In the event of any accidents the public would be reassured to know that the UKAEA would be legally liable.

Fairly soon it seems that formalised siting rules emerged. It is not known whether the same set of rules were applied to all the sites. Nevertheless, in 1955 the Reactor Location Panel (later known as the Reactor Safety Committee) was established. It consisted of members from the UKAEA, relevant government departments and the Central Electricity Authority (later to be the CEGB). The location of sites for civilian nuclear power stations were subject to the approval of this committee until the 1960 Nuclear Installations Act came into being. Under this Act a site licence was required from the Ministry of Power. The Act also established the Inspectorate of Nuclear Installations, with the responsibility of advising the Minister on all matters concerned with nuclear safety. However, the UKAEA sites are exempt from these controls since it was regarded that the 1954 Atomic Energy Act made the UKAEA responsible for their own safety.

Some reasons for the selection of sites

Whilst all the reasons for the locational decisions of the UKAEA are not known, some details relating to certain sites have emerged. The Windscale site was apparently selected to cope with the possible hazards from tanks in which highly active waste effluent from the chemical separation of plutonium was to be stored and the need for sea disposal of at least some of the liquid wastes. Some studies of tidal flows in the Irish Sea were performed but there was no detailed information on dispersal patterns and mixing characteristics before the Windscale pipeline was laid in 1950 (Danckwerts, 1983). There were powerful pragmatic reasons to go ahead and no public interest or opposition to it. The proposed effluent limits were based on extraordinarily naive analyses. It was suggested that it would be

FIGURE 3.1 *Locations of early nuclear reactors in the UK*

safe if a few selected isotopes, such as strontium in edible fish, were kept below a certain level as determined by average weekly consumption of fish by the British population (*New Scientist*, 5 January 1984).

The Calder Hall site was used because it was conveniently located close to Windscale on land owned by the UKAEA. Proximity to Windscale was needed to ease the recovery of plutonium. Despite such pragmatic reasoning, both the Windscale and Calder Hall sites happened to comply fairly precisely with the siting requirements later devised for commercial nuclear power stations. As Hinton (1969) put it:

> If you want to be kind you will say that our engineering instincts had guided us rightly; if you are less kind you will say that we had chosen the right site for the wrong reason (p. 135).

The selection process used for the Dounreay fast reactor is even more interesting.

There are two explanations for the choice of the Dounreay site. The most likely is that it was a political choice (Pocock, 1977, p. 76). Apparently, the Member of Parliament had asked that one of the planned atomic power research establishments should be built there with a view towards reducing the very high levels of unemployment. Another explanation is that a remote site was needed because of doubts about the safety of fast reactors. In a public lecture Hinton (1977) explains the choice of the Dounreay site on safety grounds. Apparently the physicists had predicted that if the fuel in a fast reactor melted, it could form a super-critical mass and explode with the force of half a ton of TNT. To reduce the effects of such an explosion the reactor was built inside a steel sphere containment. However, because of the unknowns it had been decided to search for a remote site. In 1953 sites in Galloway and the South Ayrshire coasts were examined without success, so a survey was made of the coasts of Sutherland and Caithness, where as Hinton (1977) put it, '. . . we found what we knew to be the best site at Dounreay' (p. 11). A remote site was needed despite containment because it was thought that the containment sphere would not be leak-free. If 100 per cent containment could be assured there would have been no need for a remote site.

The site evaluation process is described by Hinton (1977) as follows:

> So we assumed, generously, that there would be 1% leakage from the sphere and, dividing the country around the sites into

sectors, we counted the number of houses in each sector and calculated the number of inhabitants. To our dismay this showed that the site did not comply with the safety distances specified by the health physicists. That was easily put right; with an assumption of a 99% containment the site was unsatisfactory so we assumed, more realistically, a 99.9% containment and by doing this we established the fact that the site was perfect (p. 11).

Hinton's siting philosophy was that it was reasonable to adjust the figures, which were in any case only guesses, to support a choice that he knew from experience was the right one. As Hinton (1969) explains:

In those days we had no alternatives to pragmatism but I think the record shows that we used our judgement wisely (p. 135).

Indeed they did, but what on earth did they tell the local residents as to why the location had been chosen?

The rationale for the choice of Dounreay for the second fast breeder reactor is also interesting. Apparently the UKAEA had been anxious to build this prototype at Winfrith Heath, which was easy to reach from London, for the benefit of any potential foreign buyers. It would also demonstrate the UKAEA's confidence in its safety by constructing it near to Weymouth and Swanage (Wilson, 1971). However, the Dounreay site did have the merit of existing fuel reprocessing facilities and the undoubted asset of being in a high unemployment area, in which the Labour Government had an interest.

It seems, then, that the first generations of reactors were viewed as being suitable for existing high risk sites; for example former TNT factories and disused poison gas plants. The siting criteria that seemed to have guided these decisions were subsequently to influence the choice of civilian nuclear power plant sites. Another legacy from these early beginnings was a highly centralised research strategy, the absolute power of the UKAEA, and the strong links with government. The 1954 Act gave the UKAEA all responsibility for the research and development of reactor systems up to the point at which a prototype commercial system could be built. Burns (1967) emphasises the major difficulties that this was to cause all subsequent nuclear power programmes in the UK. The government had given the UKAEA a monopoly of selecting and developing new systems, while they were also the principal advisors to government on atomic power. A key decision during this period was the emphasis given to gas-cooled reactors – at the very time the USAEC had decided to exclude such reactors from development because of low material economy and high capital costs (McKinney, 1956).

The first civilian nuclear power programme

By 1955 the government had decided on a 10-year nuclear power programme to meet coal and energy shortages which might otherwise slow down the rate of Britain's economic growth (Pocock, 1977). Subsequent programmes were to be mainly related to growing fuel shortages (Putnam, 1954). The 1955 programme envisaged a dozen stations. The first four would be of the Calder Hall type, the next four would be larger and more advanced designs, and the last four possibly liquid cooled light water reactors of some kind. A total capacity of 1400 to 1800MW(e), which by current standards is only slightly more than a single PWR, were to be built by 1965. The electricity authorities were informed but not consulted. The idea was that the plants would be designed and built by a consortium of firms with complementary activities. Their designs would be based on the UKAEA prototypes and they would use their commercial expertise to build up a large export trade based on Britain's lead in this new technology. This plan was trebled in 1957 to 6000MW (Cmnd 8132), in part due to a panic reaction to the Suez crisis and in part due to a gross overestimate of the potential of the Calder Hall plant, although some engineers claimed otherwise (Pask and Duckworth, 1955).

Advocates of the 1957 programme appeared to believe CEGB forecasts that the power from the MAGNOX plants completed in 1962-3 would be cheaper than power from new coal fired stations (Hinton, 1967). All the stations in the expanded programme were to be of the Calder Hall type.

From 1958 onwards the UKAEA concentrated their attention on the FBR, the SGHWR, and the AGR. They were apparently planning for between 15 and 20GW of FBRs by 1986. The popular image that Britain had a world lead in the peaceful uses of atomic energy is based on two 'firsts'. In 1956 the Calder Hall plant became the world's first nuclear power station to supply power to an electricity utility, although of coure it had been built mainly to manufacture plutonium for nuclear weapons. Second, the British government had launched the world's first 10-year programme of nuclear power stations. The problem was that all 9 stations were based on the Calder Hall MAGNOX reactor which was obsolete by the time the stations worked. The new designs for the first three stations with outputs of 300MW(e) at Bradwell and Hunterston and 275MW(e) at Berkeley were in fact based on extrapolations of the prototype designs into untried regions (Pocock, 1977, p. 55). In general these reactors proved uncompetitive, and after two initial export orders they became unexportable. As Burns (1978) explains:

. . . the consortia formed in 1955-57, to design and construct the plants in this costly programme, were all engaged in the last development stages of a system with no future. This was disguised by the practice of calling the plants 'commercial' (p. 9).

These problems were compounded by the UKAEA decision in 1957 that the AGR should succeed the MAGNOX. A 30MW(e) AGR prototype began operation in 1963 the same year, as two US utilities ordered 600MW(e) Boiling Water Reactors (BWRs) as competitive plants. Yet by 1958 it was known that the cost forecasts on which the MAGNOX programme had been based were wrong and that the errors were pro-MAGNOX (Dell, 1973, p. 154). Additionally, the exponential growth in electricity demand had begun to slow, and the coal shortages of the mid-1950s were replaced by a coal surplus situation. At the same time the costs of fossil fuel power plant were falling due to the introduction of new generator sets, higher steam pressures, and cheap oil. It is not surprising then that in 1960 the timescale for the first nuclear power programme was increased and it was realised that fewer than 12 stations would be needed. Indeed it was already clear to the CEGB and the SSEB that they had been committed to the installation of uneconomic nuclear plant.

Power station siting and safety aspects

According to a former chairman of the CEGB, the real and immediate practicability of nuclear power burst upon his supply engineers without much warning with the publication of the 1955 White Paper. Before that date few of his staff had taken part in any discussions and they had been unable to disseminate any information they had because of secrecy (Brown, 1970, p. 4). Thus the first that most engineers heard of any nuclear power programme was the 1955 White Paper which stated that the construction of two nuclear stations would start in 1957, followed by two more in 1958-9, and a further four in 1960. They were to be built in the normal way and be owned and operated by the supply authority.

One of the first major problems was the identification of sites on which, according to Brown (1970), 'no work whatsoever had

95

been done' (p. 5). The problem was identified in February 1955, and two sites were needed by September. The siting criteria were new, especially the heavy foundation loads, the need for about twice as much cooling water, and the safety aspects; but on the other hand there was no need for either a large fuel supply or for ash disposal. Proximity to the grid was immensely important, indeed its existence gave maximum flexibility in the choice of sites since only relatively short lines would be necessary to link fairly remote sites into the system. The logic underlying the choice of the first sites was put as follows:

> Simple considerations of fossil fuel availability, costs and transport, and location of loads pointed towards the south of the country for the first nuclear stations, but the time factor militated against looking everywhere. The question therefore was where in the south? Inland water was scarce, hence coastal siting was probable but most of the south coast was heavily built up and used for recreational purposes. In this hasty evolution the existence of the grid and supergrid was of immense value . . . (Brown, 1970, p. 5).

Another problem was that the siting requirements themselves were subject to change with increasing knowledge of the extent to which the civil stations would differ from the military Calder Hall plant in output capacity, physical features, and constructional requirements (Usher, 1962).

By March 1955, within one month of the White Paper, the site search had been concentrated on the Severn Estuary and the Essex coast; the former because of the presence of tidal water well inland, and the latter because of the moderate transmission distances to London. Aerial surveys in March and April resulted in the ground inspection of 24 sites in April, and the first trial borings in May. Concurrently there were preliminary discussions in confidence with the planning authorities and by September 6 sites had been selected as potentially suitable. By October, Bradwell and Berkeley had been chosen for the first two stations and publicly announced. Formal applications were made in December and statutory consents issued in July 1956. In 1957 the second group of two sites were selected; the first at Hinkley Point and the second to meet the expanding load in NW England, an area of high coal cost. Five sites were investigated and Trawsfynydd finally chosen. This latter site is unique for two reasons: first, it is situated on a lake in a national park; and second, it attracted a demonstration in support, because it would alleviate local unemployment problems resulting from the closure of a military camp on the site.

The early haste in the search for sites has been replaced by a far more lengthy process, some details of which were described in Chapter 2. This was necessary because the later stations were considerably larger than the earlier ones with greatly increased cooling water requirements and needed more extensive and specialised investigations. Open coast sites such as Sizewell which are exposed to the full force of gales, with waves having a long fetch, posed new problems in the siting and design of the cooling water intakes and outfall works. The siting problem was also seen as a major one because it was being forecast by the CEGB in 1961 that '. . . during the next 15 years or so, it would be necessary to find about 40 to 45 sites in Britain, each to carry generating plant of between 1000 and 1500MW output' (Gammon and Pedgrift, 1962, p. 232).

TABLE 3.1 *Decision dates for MAGNOX reactors*

Site	Reactor location panel or reactor safety committee approval	Section 2 consent	Nuclear site licence consent	Size MWe	Date of operation
Berkeley	22.9.1955	18.7.1956	30.3.1960	137	1962
Bradwell	22.9.1955	10.7.1956	30.3.1960	150	1962
Calder Hall	1953?			60	1956
Chapelcross	1955?			60	1959
Dungeness	5.11.1957	12.8.1959	14.10.1960	275	1966
Hinkley Point	22.9.1955	25.6.1957	30.3.1960	250	1965
Hunterston	1956?			150	1956
Oldbury	17.6.1959	20.10.1960	20.12.1960	280	1967
Sizewell	21.11.1957	2.2.1960	21.4.1961	290	1966
Trawsfynydd	26.7.1957	1.8.1958	30.3.1960	250	1965
Wylfa	17.9.1959	29.12.1961	25.11.1963	590	1971

Source: Adams and Faux (1969).

Table 3.1 gives the dates of critical decisions for the MAGNOX stations. The Hinkley Point site is particularly interesting. The first positive public information about a station here appeared in a local newspaper on 27.11.1956, over a year after the CEGB had sought planning approval from the reactor location panel. There was eventually a public inquiry which lasted 2 days and the Minister of Power gave his consent, emphasising the difficulty of finding such suitable sites; the CEGB had emphasised the firm rock foundations and the abundant supplies of cooling water.

It is also interesting that all the MAGNOX stations in Table

3.1 were sited at a time when there was no legislation dealing specifically with the siting and operation of civilian nuclear power stations. In practice this was not a problem because no station was actually in operation until 1962 when the Nuclear Installations Act of 1960 was in force. Finally, it is worth noting that all nine of the CEGB and SSEB stations can be regarded as prototypes in the sense that each incorporates some significant technical change in components. Despite this the 1955 programme was a success. It was designed to reduce coal consumption to a relatively constant level and this it did. The main problem was that the nuclear industry was not well equipped to meet either the drastic changes in political attitudes to nuclear power or to reorganise its strategy to meet fuel policies that were being altered in less time than was needed to build a power station. In the early 1960s it took 5 years to build a MAGNOX station whilst on average national fuel policies changed once every 3 years!

Other problems concerned whether public safety was being given sufficient attention during these early years. The atomic stations were being built by firms who had no relevant experience and they were to be run by Central Electricity Authority (later the CEGB) staff who had no knowledge of nucleonics. The creation of the CEGB in 1958 with Hinton (formerly director of UKAEA Risley) as its first chairman, started a pronuclear trend that has continued ever since. Hinton was considered to be the architect of commercial atomic power production and had been pressing the issue since at least 1946.

In general it seemed that in the 1950s little attention was given to safety aspects and the dangers do not seem to have been openly discussed. The engineers were so absorbed with the new technology that accidents were confidently never expected to occur. Many of the early designs look now to have been a distinct gamble. At the time of the 1955 White Paper there was little information available on the irradiation behaviour of materials and virtually none on complete fuel elements (Williams, 1980, p. 41). The MAGNOX fuel cans had not been perfected when the first commercial stations were being ordered. It is perhaps slightly miraculous that nothing serious occurred during the period when experience and reliability data were being built up. The principal safety measure was the use of a 'remote' siting apart from the usual design aspects. The problem was of course that no one had any real idea as to what an accident might involve and how remote a 'remote' site needed to be. It was all a matter of judgement and, since there were no nuclear disasters, it could well be claimed that the correct decisions were made; but it was certainly a gamble with only a vague idea of the public safety stakes.

So great was the rush to develop commercial nuclear power that the first 5 stations received their nuclear site licences after construction was well advanced (Adams and Faux, 1969); see Table 3.1. The Windscale site was exempt. The principal hazard here was seen to be the storage of the high level waste effluent from the chemical separation of plutonium. At that time there were no hazards thought to be associated with the reactors (Hinton, 1969, p. 135).

Additionally, the heavy engineering industry was expected to build up from scratch the teams of experts needed to build atomic power stations. As Toombs (1977) recalls, 'Such a collection of disciplines under a single control was new to the heavy plant industry' (p. 5). The design teams greatly increased the heat from each tonne of nuclear fuel. The 1955 White Paper had envisaged a reactor size of 75MW but even the very first designs were for 150MW reactors. The speed of change was absolutely astonishing. Toombs (1977) reports that

> The consortia then set about preparing their tenders in great detail, backing each item of novel design with analytical and experimental evidence. The tenders were voluminous documents weighing many hundredweights. They were submitted in October 1956. The assessments were completed and the first three contracts awarded in December 1956 (p. 6).

As Pocock (1977) points out the technical examination of these proposals by the electricity authorities and their consultants *took all of November*!

Remote siting policies

The 1955 White Paper stated the basic principle that '. . . the first stations, even though they will be of inherently safe design, will not be built in heavily built-up areas' (Cmnd 9389, 1955, p. 9). This negative definition of remoteness as not being in heavily built-up areas is a very characteristic feature of reactor siting policies in the UK. The term 'remote' is not what would normally be interpreted as being geographically isolated. It was always intended that practical nuclear power station sites would be required near the major electricity load centres and the problem then, as now, is how to select suitable sites with an adequate degree of remoteness in regions of relatively high population density and where sizable towns are widely distributed. To assist this process a number of demographic siting criteria have

99

evolved, primarily as a means of reducing the public health risks associated with nuclear power under accident conditions. It has always been maintained that these risks are minute and that the avoidance of large population concentrations was dictated by prudence, indeed there was no explicit scientific basis for the population and distance limits that were applied.

Marley and Fry's siting criteria

The choice of sites for the first generation of civilian reactors was determined not only by engineering requirements but also by factors connected with the nuclear side of the plants in normal, and under possible accident conditions. Particular importance was attached to fears of the possible effects of a release of fission products on the surrounding populations although it is doubtful whether anyone had any clear idea of what a reactor accident might entail. There was only limited data about the release of fission products from fuel and little data about deposition rates or the effects of the radioactive material on people via ingestion or inhalation (Farmer, 1979, p. 262). Parker and Healy (1955) assessed the consequences of a major reactor accident as leading to the deaths of 200 to 500 people, in an area where these numbers also represented the population density per square mile. Others thought, wrongly as it happens, that fission products presented a new hazard of between 10^6 and 10^9 times more dangerous than chemical hazards (McCullough, 1955). So although the first reactors were very small by current standards, there was considerable concern for the possible effects of a large release of radioactivity from a reactor accident (Mesler and Widdoes, 1954). One wonders whether this was either because major accidents had already occurred at isolated military reactors in the USA or because it was easy to measure ground and airborne activity rates from even small releases of fission products. Many of the early research reactors were air-cooled – for example, those at Harwell – and various interesting empirical experiments could be carried out on the diffusion of fission products over neighbouring regions (Stewart *et al.*, 1954).

Marley and Fry (1955) describe the development of techniques for estimating the radiological hazards resulting from the release of between 1 and 10 per cent of the fission products of a 250MW power station. In the absence of any practical information about the nature of real reactor accidents, they studied the problem from the standpoint of various notional releases of activity. The best they could manage was to list various direct and indirect effects, scaled relative to the proportion of fission products formed inside a reactor. To provide an indication of possible geographical effects they estimate the maximum range downwind to which various levels of ground contamination might be expected to extend. They suggest that the downwind range, R, is roughly proportional to the square root of the amount of activity released resulting in the formula

$$R = B \times M^{0.5} \tag{1}$$

where
M is the power of the reactor in megawatts multiplied by the fractional release,
B is a constant that varies according to atmospheric conditions and the limit of ground level contamination considered relevant – for example, a dry weather value of 18 would be used to indicate the probable limit of temporary evacuation (Marley and Fry, 1955, p. 104).
Under conditions of rain, these distances may be increased by a factor of up to 2.8 at distances over 3 miles. Under inversion conditions, the formula does not apply and the ranges may be larger than indicated by equation (1).

TABLE 3.2 *Estimated maximum hazard ranges under dry turbulent conditions according to Marley and Fry (1955).*

| | Range in miles | | |
| | Size of release | | |
Effect	0.1MW	0.2MW	1.0MW
Urgent evacuation within 12 hours	.18	.31	.68
Evacuation or restrictions on normal living	1.0	1.43	3.23
Probable limit of temporary evacuation	3.54	4.97	11.19
Probable limit of temporary milk contamination	11.75	16.66	37.29

The Marley and Fry formula relates to releases of 1kW to 1MW of reactor power. For a Calder Hall size of MAGNOX reactor this would be equivalent to a fractional release of less than 1 per cent. Table 3.2 shows the geographical ranges of various consequences for a fission product release under dry turbulent atmospheric conditions. By reference to tables of this type it would be possible to derive population distributions that

101

would minimise the consequences of a release of an assumed size. Marley and Fry (1955) conclude

. . . that the release of a significant fraction of the fission products accumulated in a high-power reactor cannot be contemplated within many miles of a normally inhabited area (p. 104).

On the other hand, as Table 3.2 shows a release of fission products restricted to 0.1MW of reactor power should not necessitate lengthy evacuations beyond a narrow sector extending 1 mile downwind, temporary evacuation might be needed out to 3.5 miles and contaminated milk out to 11.7 miles. However, if accidents could be restricted to involve releases of no more than a few tenths of a megawatt equivalent of fission products, then according to Marley and Fry remote siting should be unnecessary.

TABLE 3.3 *Remote siting criteria according to Charlesworth and Gronow (1967)*

Distance band (miles)	Population limits in any 10 degree sector	All-round population limits
0-0.3	few (50)	few (300)
0.3-1.5	500	3,000
1.5-5.0	10,000	60,000
5.0-10.0	100,000	600,000

It is generally claimed that the remote siting criteria used to locate all the MAGNOX reactors and the first two AGRs are based on Marley and Fry (1955); see for example, Charlesworth and Gronow (1967). The population siting criteria generally attributed to Marley and Fry are given in Table 3.3.

The problem is that Marley and Fry (1955) do not specify any such limits. The answer seems to be contained in an unpublished and still restricted report by Fry (1955); the UKAEA still refuse access on the grounds that it contains sensitive information (letter to the author in 1984). According to Charlesworth and Gronow (1967), Fry (1955) argues that very few people should be exposed to extreme risks and urgent evacuation plans would be needed for these people. In addition protracted evacuation or severe restrictions on normal living should not be imposed on any but small population centres. Temporary evacuation should not be necessary for more than 10,000 people except under exceptional weather conditions, and if an accident were to coincide with exceptional weather conditions no more than 100,000 people should be affected. The link with Marley and Fry (1955) is simply that the critical distances in Table 3.3 can be matched against the hazard ranges for various consequences for a 0.2MW release in Table 3.2. The fit is good enough to explain the first three distance bands. The fourth one might have been expected to have been 2.8 times larger than that shown in Table 3.2; this being the factor that Marley and Fry used to represent the worst weather conditions (rain). However, this factor of 2.8 might well be regarded as a maximum value and if a value of 2 is used instead then the equivalences between the distances in Tables 3.2 and 3.3 are more accurate.

The rationale for the 10 degree sectors in Table 3.3 is that the atmospheric dispersion of a short term release of fission products over a period of minutes rather than hours might be expected to be restricted to a 10 degree horizontal arc. The more usual $22\frac{1}{2}$ and 30 degree arcs are more appropriate for releases extending over several hours. The all-round population limits based on six times the 10 degree sector limits are arbitrary. They are basically a device to incorporate population density in the region of the reactor.

The only remaining problem is how the population limits were obtained. Again the answer is probably in the unavailable Fry (1955) report. According to Bell and Charlesworth (1963), Fry (1955) contains details of a study of the population distribution around reactor sites in the UK. The sites were classified in a geometric progression, the classification depending on the distances within which populations of 500, 10,000, and 100,000 were contained. These groups were defined by considerations of evacuation or other restrictions which might arise from the deposited activity from a mixed fission product release. The population limits themselves could well have been obtained in the following manner. Outside of the major urban areas the population density in Britain is about 10000 persons per square mile, according to information available in the 1950s. For any 10 degree sector this would give approximately 87,000 within 10 miles, rounding up you get 100,000. For the 5- to 10-mile distance band the average density gives a population limit of 10,908 or 10,000 by rounding down. For the 1.5 to 5 mile band you get a population of 491 or 500 if you round up. Once again it seems that a little detective work produces results very close to the actual values and the logic used here probably gives an indication of what lay behind the remote siting criteria.

Farmer and Fletcher's criteria

By the late 1950s the situation had changed following the Windscale plutonium pile accident in 1957. In particular, experience with the 'large' reactors at Calder Hall had made it possible to obtain more realistic estimates of the likely total release of fission products. This led to the view that any releases would mainly consist of volatile fission products principally iodine 131. Under credible accident conditions it was thought likely that only small releases would occur from the Calder Hall type of reactors (Brown *et al.*, 1958). The most pessimistic accident that could be imagined, and thus credible, was thought to involve a release of 250 curies of total iodine or 10 curies of iodine 131, the latter being the principal health hazard during the Windscale accident.

Farmer and Fletcher (1959) report the results of a comparison of what they refer to as the 'existing siting criteria' derived from the notional release of mixed fission products as described in Marley and Fry (1955), with their modifications to the criteria which would result if only total iodine were to be released. They also argued that it would be prudent to allow for additional factors of safety by assuming that either the release of activity takes place under unfavourable conditions of wind direction and moderate temperature inversion, or that the release is ten times larger than the figure assumed from a pessimistic accident analysis. Medical effects were based on the latest recommendations of the Medical Research Council and took into account the effects of inhalation, consumption of exposed food, and whole body radiation from the ground and the cloud. The hazard ranges Farmer and Fletcher obtained are shown in Table 3.4.

Table 3.4 suggests that beyond about 1 mile there need be no restrictions on population distributions although extensive monitoring may be needed out to at least 2 miles and perhaps 5 in order to give assurances about safety. These distances bear a very close resemblance to the current emergency evacuation plans for UK power stations (HSE, 1982b). Farmer and Fletcher (1959) also advise that there should be few people living within .25 miles and that within 1 mile the population should be mobile and not include large numbers of children, that there should be no open water reservoirs, no photographic industries, and no airfields; finally, the site should have good access for emergency services. The aim then is basically to find sites that have few people living in close vicinity, a feature which was later to become the only important criterion.

According to Bell and Charlesworth (1963), the Marley and

TABLE 3.4 *Estimated maximum hazard ranges (miles) according to Farmer and Fletcher (1959)*

Effect	Fair weather		Moderate inversion
	250 curies released	2500 curies released	250 curies released
Urgent evacuation	.06	.06	.06
Temporary evacuation	.25	.87	.49
Temporary ban on contaminated food	.31	1.00	.62
Temporary ban on milk consumption	3.97	13.98	8.94

TABLE 3.5 *Remote siting criteria according to Farmer and Fletcher (1959), and their modifications to it*

Distance band (miles)	Population limits
Current criteria	
0-0.28	Few (50)
0.28-1.5	Less than 500 people in any 10 degree sector
1.5-5.0	No large centre of population of 10,000 people or more
Revised criteria	
0-0.25	Few (50)
0.25-1.0	Less than 500 people in any 30 degree sector
1.0-5.0	No large centre of population of 10,000 people or more

Fry (1955) criteria were replaced by the Farmer-Fletcher (1959) criteria. This is interesting because Farmer and Fletcher do not explicitly list any. Furthermore, their understanding of the 1955 criteria appears to be different from that shown in Table 3.3. Table 3.5 gives the old and the revised siting criteria that are quoted in Farmer and Fletcher (1959). The 5 to 10 mile distance band seems to have been dropped.

Farmer's criteria (Farmer, 1962)

The basic remote siting criteria were further improved by Farmer (1962, 1962b). Although the improvements came too late to have any direct effect on siting decisions, they are of interest because they were almost certainly used to identify some later sites when applied in a relaxed manner, and they were used to control population developments around existing sites.

A major criticism of the Marley and Fry (1955) and also of the Farmer and Fletcher (1959) criteria is that it was difficult to make comparisons between sites with a number of small close-in developments and those with large more distant towns. A site 4.9 miles from a large town might not be acceptable whilst a site 5.1 miles from a substantially larger population centre might be. It was thought desirable, therefore, that a more refined approach would be required to allow a fair comparison to be made between sites around which the population is distributed in different ways. Indeed the concern was not with maximising public safety only with a means of comparing sites with respect to their population distributions in the neighbouring regions.

Farmer (1962) argued that there is no need to seek sites which protect the public against normal reactor operation, since any operational releases are both strictly controlled and harmless. Instead sites would be evaluated by reference to purely hypothetical releases of fission products. By 1962 experimental work on the releases of fission products from fuel elements and from studies of actual accidents had suggested that the dominant hazard to a resident population would arise through the inhalation of iodine 131. Thus sites should be assessed in relation to this factor. It is noted that this implies a concern with prompt or short-term protection measures and that the longer term consequences of, for example, caesium 137 land contamination, are ignored as irrelevant to the site evaluation process.

The downwind concentration of airborne fission products depends on distance and the prevailing meteorological conditions. Pasquill (1961) provides a basic atmospheric diffusion model which could be used to predict ground-level air concentrations of fission products under a number of different atmospheric stability categories. Farmer selected Pasquill category F as an extremely unfavourable condition although not necessarily the most hazardous, on which to base his development of a siting criteria. Pasquill category F weather is characterised by low wind speed (about 2 m/s) and a clear night with a moderate temperature inversion. This combination of circumstances would slow down the diffusion processes and greatly increase airborne concentrations. Such weather is thought to occur with a frequency of between 5 and 20 per cent.

In 1962 it was not considered feasible to try and obtain an absolute value for the hazard to which people living at different distances around a reactor are exposed. At that time there, was no means of predicting either accident frequencies or likely sizes of releases, little precise information was known about the biological consequences of exposure, or indeed about the nature and composition of the released activity. So Farmer (1962) considered the case when only 1 curie of radiation was released at ground level under Pasquill category F weather. He assumed that the cloud dose was uniform across a 30 degree sector to allow for some variation in wind direction whilst the release was occurring. He then obtained estimates of the ground level concentration of fission products for a single 30 degree sector at various distances. It was thought that people living near to the reactor might be affected by fission product releases of large or moderate size under a variety of atmospheric conditions, whilst those at greater distances would only suffer adverse consequences during unfavourable weather conditions. Farmer allows for this by squaring the predicted airborne concentrations.

This approach results in a series of weights which represent the estimated squared airborne concentrations of fission products resulting from the 1 curie ground level release; see Table 3.6. An assessment could now be made for any site by dividing the population for each 30 degree sector into 1 mile distance bands. The resulting populations are multiplied by the weights shown in Table 3.6 and divided by 1000 to yield a site rating factor; the maximum value for any sector would be used to characterise the site in question. Thus the site rating factor for the jth 30 degree sector can be computed as

$$S_j = \frac{\sum_{i}^{n} P_{ij} W_i}{1000} \quad (2)$$

where
n is the number of distance bands used;
P_{ij} is the population in the jth 30 degree sector for the ith distance band from the reactor site; and
W_i is the weighting factor for the ith distance band, see Table 3.6.
The maximum S_j [j = 1,2,. . .12] value is used to represent the site.

107

TABLE 3.6 *Site Weighting Factors*

Distance band (miles)	Weighting factor
1-2	259
2-3	48
3-4	18
4-5	9.6
5-6	5.3
6-7	3.6
7-8	2.7
8-9	2.0
9-10	1.4
10-11	1.1
11-12	0.9
12-13	0.760
13-14	0.626
14-15	0.524
15-16	0.444
16-17	0.381
17-18	0.330
18-19	0.288
19-20	0.254

Source: Farmer (1962).
Values for 13-20 miles estimated by author.

The resulting site rating can be categorised as follows. Farmer argued that the maximum possible site rating is about 12,000; this being obtained for a hypothetical site with population totals equivalent to the highest values found in the UK for any urban area (about 40,000 persons per square mile). From this he derives four classes of site: (1) a class I site with a rating of up to 750, this would be equivalent to about 6 per cent of the maximum population density; (2) a class II site with a rating of 751 to 1500; (3) a class III site with a rating of 1501 to 3000; and (4) a class IV site with a rating of 3001 to 6000.

This typology was justified on the grounds that site ratings for sites which had already been approved for power reactors had ratings of up to 750. One problem is that the precise value obtained from equation (2) depends on the number of distance bands used in the summation. The lower distance band of 1 mile is fixed by the need to ensure that there is a highly mobile and low population in this area; see Farmer and Fletcher (1959). For practical purposes the outer summation distance is fixed such that a town of 0.5 million people will not change the class allocation of a site. Thus a limit of 20 miles is applied to class I sites, 12 to class II sites, 6 to class III sites and 4 to class IV sites. These variable summation distances present some difficulties because the published weighting factors are only given for 1 mile distance bands out to 12 miles which is insufficient to properly assess a class I site. This problem was resolved by replicating the procedure used by Farmer to compute the weighting factors and then extend the distances out to 20 miles.

One problem with these Farmer remote siting criteria concerns the use of fixed 30 degree sectors with an arbitrary origin. It has been suggested that the use of 30 degree sectors enables most communities to be considered in a single 30 degree sector, whilst any narrower sector width could lead to the unrealistic division of urban areas. Another query might well concern the use of 1 mile wide distance bands. The use of a finer distance division close to the reactor might well make a difference to the site rating. Indeed Bell and Charlesworth (1963) give a very fine 0.25 mile breakdown for areas within 3 miles which is presumably the limit to which accurate population counts could be obtained at the time.

Farmer emphasises that the site rating is only a guide. Other factors have to be considered; for example, the overall distribution of population, the local topography, the relative prevalence of different wind directions under inversion conditions and their persistence. Many of these factors do not lend themselves to simple numerical presentation and their consideration must be a matter of judgement. Likewise the choice of an appropriate class for a particular reactor system requires an informed technical judgement.

Criticisms

In the UK the 'remote' siting criteria were never meant to make full use of geographical isolation as a safety measure but merely to avoid locations within 20 miles of large towns. The criteria that evolved were largely arbitrary in that they bore little or no relationship to the likely effects of reactor accidents and they could well have been designed or calibrated to reflect the population characteristics of the existing UKAEA reactor sites. Another problem concerns the manner in which population estimates were obtained for segments of circles at a time when there was no readily available or accurate local population data. The only way accurate population counts could be obtained was by counting houses and then making some assumption about average household sizes; this would only be practicable for areas close to a site that had been short-listed for more detailed investigation; it clearly could not be done during the site search phase. Even then it seems that most of the population distributions were based on average densities computed for local authority areas assuming a uniform distribution within a region using data obtained from the 1951 census and Whitakers' Almanac (private correspondence with Mr Tweedle, Document Classification Officer, UKAEA Harwell, 1980, concerning Fry (1955)). These data problems would be made worse by the use of an arbitrary origin for the 10, 30, and 60 degree sectors that were used. A town bisected by a sector boundary, or a distance band boundary, may easily result in an acceptable site, whereas if the sector origin had been set at any number other than 0 degrees the site may have been flagged as violating the criteria. A simple solution is to rotate the sectors through increments of 1 degree starting at 0 but whilst this is easy for a computer it would have been impossible for the manual methods used in the 1960s.

Finally, it is not clear as to how any siting criteria are applied in practice. Bell and Charlesworth (1963) wrote:

> . . . the need for some basic rules to be applied in the initial selection of sites has been amply demonstrated. It is also clear that it is neither possible nor useful to attempt to define in advance all the conditions which enter into a site evaluation. Some sites, which appear to be marginally unacceptable, may turn out on closer inspection to have compensating advantages whilst perhaps others, although conforming with the laid-down requirements, may possess such unfavourable characteristics when viewed more broadly that the site is rejected (p. 325).

This is a feature of all siting policies, although it becomes unsatisfactory if the siting criteria themselves are subject to negotiation; for example, site 'x' is a good site, what a shame it marginally violates the demographic criteria! There is no evidence that this has happened with the remote siting criteria although it could be argued that given the inaccuracies of the available data that any slightly marginal site in relation to the criteria would almost certainly violate it. Indeed two of the MAGNOX sites were considered to be distinctly marginal. In general no documentary material survives to provide details of what actually happened in practice. Researchers who were employed to perform the site searches are reticent about discussing details of their procedures, as they are still covered by the Official Secrets Act. However, there is nothing to stop a dedicated geographer from trying to replicate the various remote siting criteria and then applying them retrospectively to the existing nuclear power station sites using the best available data sources. The fact that 1971 or 1981 census data are to be used to examine sites selected in the 1950s is of no real consequence because it has been often stated that planning controls on new developments have been used to preserve the favourable demographic features of the remote sites. The principal problem is determining accurate coordinate locations for the sites. The CEGB initially declared that this information was secret even though most sites are marked on 1:50,000 Ordnance Survey maps. Generally it is not possible to distinguish between A and B stations or between individual reactors. Nevertheless the coordinates used here were found to closely match those subsequently provided by the CEGB.

Application of the remote siting criteria to UK sites

All the siting criteria described here require population estimates for various circular- and wedge-shaped zones. In the UK previous studies seem to have relied on poor quality, inaccurate, but readily available data. There have been suggestions that some of these data seem to be grossly inaccurate (Openshaw, 1982b), indeed it was only after 1976 that population counts for spatially referenced small areas were available for the whole of the UK in computer readable form. It is possible that some of the existing sites may actually violate the siting criteria on which they were selected, although census data are themselves too coarse to be certain. Chapter 7 provides a full discussion of the problems of demographic estimation relevant to nuclear siting criteria.

Data errors apart, it also seems that the recent availability of national demographic data for the finest resolution census units has meant that it is unlikely that the full range of feasible sites has ever been identified with any precision. The tremendous advantage of demographic siting criteria is that they can be automated and used to identify all potential sites. In the UK there is no evidence to indicate that such extensive searches have ever been performed by computer, which is common practice in the US, as distinct from by manual methods. Indeed it seems that the CEGB take considerable pride in the fact that computer searches of this type are not used; the people involved seem to regard the human touch as being necessary.

Table 3.7 presents the results of applying the three versions of the Marley and Fry (1955) criteria to all the existing sites. Population data from the 1971 and 1981 census are used mainly because of the different estimation methods used for each data

111

TABLE 3.7 *Early remote siting criteria applied to existing UK reactor sites: sites which violate the criteria are starred*

site	Charlesworth and Gronow			Farmer and Fletcher			revised Farmer and Fletcher		
	a	b	c	a	b	c	a	b	c
UKAEA sites									
Chapelcross				**	*	*	*	*	*
Dounreay									
Harwell				**	**	**	**	**	**
Windscale				**	**	**	*	*	*
Winfrith				**	**	*	*	**	*
Remote sites									
Berkeley				**	**	*	*	*	*
Bradwell					**	**	*	*	*
Dungeness									
Hinkley Point									
Hunterston				*	*	*	*		*
Oldbury				**	**	**	**	**	*
Sizewell				*	*	*	*		*
Trawsfynydd									
Wylfa									
Relaxed sites									
Hartlepool	**	**	**	**	**	**	**	**	**
Heysham	**	**	**	**	**	**	**	**	**
Torness									
Druridge Bay				**	**	**	**	**	**

Notes: a based on 1971 1km grid-square data
b based on 1981 1km grid-square data
c based on 1981 census enumeration district data
*site violates the 5-mile population restrictions
**site violates the 5-mile population restrictions and one or more others

set. Differences in results may be regarded as being due to population growth and different spatial resolutions of the data; the 1971 data are based on 1km grid-squares, the 1981 on census enumeration districts. The results show that the Bell and Charlesworth (1963) version of the remote siting criteria are satisfied by all the sites except Hartlepool and Heysham which were selected using very different criteria. However, there is a real problem in defining what is meant by the term 'few'. If 'few' is quantified as indicated in Table 3.3 then 4 of the remote sites would violate this constraint in 1971 and 5 in 1981 (Openshaw, 1984) – although it is possible that these violations may be due to data estimation problems. It may also be that this is where an element of judgement enters into the site evaluation process since this close-in region would be where prompt evacuation would be needed.

The situation with respect to the other two versions of the remote siting criteria is far more dire. Most sites violate these criteria even if the close-in populations are ignored. It seems that the 5-mile distance bands are readily exceeded and that some sites also fail by having too great an all-round population. Perhaps too much emphasis should not be placed on the significance of these violations because of uncertainty as to whether they were actually used. Attempts to clear up this problem by asking the UKAEA have so far failed. Nevertheless, the results are interesting if only as an indication of the seemingly retrospective attempt to devise a consistent set of criteria. Remember the ultimate cause of these uncertainties is the fact that the remote siting criteria are attributed to Marley and Fry (1955), who present no such criteria only a basis for developing some.

Table 3.8 shows the results of applying Farmer's revised remote siting criteria (Farmer, 1962) which were in fact never used as all the MAGNOX stations and the first three AGR's were sited according to the Marley and Fry (1955) criteria (Charlesworth and Gronow, 1967). The criteria are implemented here as devised by Farmer with fixed 30 degree sectors. These results are interesting in that Sizewell appears to violate these criteria with all three data sets that were used. Whether this is significant or not can be questioned on the grounds that the Farmer siting criteria give most weight to close-in populations (see Table 3.6), and again it is here that demographic estimation errors will be greatest. This sensitivity to what may be termed data uncertainties probably explains the variation in some of the results for particular sites between 1971 and 1981. It may also indicate that sites which exhibit such a high degree of data sensitivity may well be more marginal than the siting score would indicate.

Using the computer to search for feasible sites in terms of demographic criteria

The results in Tables 3.7 and 3.8 give no impression of the extent to which the remote siting criteria may have excluded large parts of the UK from consideration as nuclear reactor sites. The impossibility of automated searches in the 1950s through to the late 1970s may have allowed the CEGB to avoid the fairly unpleasant business of having to evaluate large numbers of

TABLE 3.8 *Later remote siting criteria applied to UK nuclear reactor sites, Farmer's site ratings*

Site	1971 1km data site rating	1981 1km data site rating	1981 ed data site rating
UKAEA sites			
Chapelcross	616 (I)	520 (I)	585 (I)
Dounreay	34 (I)	44 (I)	72 (I)
Harwell	484 (I)	435 (I)	649 (I)
Windscale	372 (I)	401 (I)	436 (I)
Winfrith	273 (I)	361 (I)	640 (I)
Remote sites			
Berkeley	311 (I)	328 (I)	366 (I)
Bradwell	429 (I)	272 (I)	302 (I)
Dungeness	85 (I)	92 (I)	114 (I)
Hinkley Point	207 (I)	119 (I)	113 (I)
Hunterston	358 (I)	386 (I)	333 (I)
Oldbury	323 (I)	348 (I)	409 (I)
Sizewell	819 (II)	810 (II)	778 (II)
Trawsfynydd	220 (I)	100 (I)	95 (I)
Wylfa	455 (I)	256 (I)	379 (I)
Relaxed sites			
Druridge Bay	618 (I)	988 (II)	1088 (II)
Hartlepool	1499 (II)	1225 (II)	1299 (II)
Heysham	1926 (III)	1937 (III)	2496 (III)
Torness	101 (I)	51 (I)	108 (I)

Note: site ratings based on fixed sectors, the values can all be increased by the use of optimal rotations.

alternative sites. The manual search procedures seem to have found so few potential sites, that each suitable site is so rare and precious that it has to be safeguarded for development in the national interest. Whether this folklore is accurate depends on whether there are really so few sites, and this can only be determined by automated searches for sites over the entire UK land area.

In the USA a number of strategic studies have been made of the impact of various siting criteria in terms of the areas of states that would be excluded; see Chapter 6, as well as Aldrich *et al.* (1982), Durfee and Coleman (1983), and Kelly *et al.* (1984). These US studies have typically used a grid of 5km by 5km to examine the whole of continental United States. The feasibility of locating a nuclear power station at the midpoint of each cell can then be assessed and the areas that would be excluded by various criteria computed. To guide the interpretation of such analyses Kelly *et al.* (1984) have put forward the following arbitrary rules of thumb. A state with a minimal nuclear siting problem is defined as one for which at least 67 per cent of the initial land area is available for siting nuclear plant, or over 50 per cent of the total area available. They suggest that 'No siting problems are likely to occur in these states, no matter which demographic restrictions are chosen' (p. 22). Similar automated geographical site searches have been applied to the UK – see Openshaw (1980), (1982a). In these UK studies a spatial resolution of 1km was used and the results for both a coastline corridor and the total land area reported. The intention here is to use a more sophisticated version of these search methods to allow a more accurate determination of exclusion areas to be computed. Two criticisms can be levelled at the previous UK work: (1) the maps exaggerate the areas of infeasible sites due to problems with scaling the plotter line width – on the maps presented in Openshaw (1982a) the scaled pen width was 3.2km instead of the intended 1km; and (2) no allowance was made for infeasible sites which were located outside the UK land area. The search was over a rectangle of 1220 by 655 km in 1km increments and counts were made of the numbers of infeasible grid-squares with respect to various demographic criteria. This is of course quite reasonable if floating and, or, offshore sites are to be included; indeed such sites have been suggested in the US. However, it would seem desirable to adjust the area counts to refer only to the UK land area.

It is by no means an easy task to refine these area counts so that they refer only to the UK land area. The only practicable approach is to classify all the 1km grid-squares that cover the UK in terms of whether or not they lie within the digitised outlines of the boundaries of each country in England and Wales or regions in Scotland. A point in polygon technique was used and those 1km squares which have 3 or more key points (defined as the 4 corners plus a centroid) within a particular county's polygon were allocated to the area concerned. It is interesting that the resulting county and region polygon-grid square approximations resulted in an error of only 70 km^2 in estimating the total UK land area; the maximum county area error amounted to less than 1.5 per cent. Once this grid-square to county allocation process had been performed it was easy to make accurate counts of the areas of invalid sites both for Britain as a whole as well as for individual counties and regions within it.

A similar approach was used for the UK coastal corridor search. The digital outline of the UK as supplied by the

Ordnance Survey was used to identify all those 1km squares which were intersected by the UK coastline. The resulting 13,614 coastal grid-squares could then be examined for feasibility as nuclear power station sites. This has the advantage of being more realistic than a total area search; indeed all but one of the current civilian nuclear power stations are located on coastal or estuarine sites which are included within this coastal corridor. It provides a useful complement to the whole area searches. Once again the coastal grid-squares can be allocated to the county or region of which they are part. This task was made a little difficult because the coastal outline coordinates were far more accurate than the county and region polygon boundaries. The latter had to be regarded as fuzzy regions with an average fuzzy factor of 2 km.

The only remaining task is to describe how the search of all 799,100 1km grid-squares within the search rectangle was to be performed. This analysis is based on 1971 census 1km grid-square population counts for some 150,000 cells. Large amounts of computer time are required for these searches and the analysis was only possible through the development and use of a highly efficient spatial data retrieval method for use with grid-square data (Openshaw, 1982a). These techniques reduced estimated computer run times on an IBM 370/168 from several days to a small number of hours, typically between 1 and 3. The results could then be mapped with a resolution of 1km or reported for counties, regions, or all of Britain. It is suggested that this style of automated geography can be taken and developed further (Dobson, 1983) and provides a very effective way of evaluating the impact of alternative siting policies and for searching for alternative sites to be the subject of a subsequent and more detailed evaluation. This sort of approach is now regarded as the standard way of approaching the siting problem in the US, and its use in the UK by the CEGB is to be highly recommended. It is no longer sufficient for the CEGB to take the view that their nuclear stations are so safe that extensive searches for the safest sites are not necessary. Nor is it good enough to argue for the existence of such rigid engineering constraints that the areas of site search are so geographically localised that it is scarcely possible to find a single site let alone a few alternatives for evaluation purposes. In the past this strategy has worked because siting was not regarded as a contentious issue, the computer tools were not available, and site selection was regarded as an urgent process to be performed within a very short time scale. None of these conditions apply today.

It may be that the informal judgement of an experienced planning engineer is still the best means of identifying the best sites from the CEGB's point of view but this is no longer sufficient, nor would it be in the US, to determine the siting decision. It is necessary to 'prove' the case by identifying all possible potential alternatives and then demonstrate the superiority of the preferred site in an explicit manner; it cannot simply be assumed. Once the site search process is able to consider all potential sites not excluded on absolute technical and demographic grounds then considerations of public safety can enter directly into the site evaluation and comparison process. Until the CEGB start doing this then they can be fairly criticised for failing to do their job properly.

To forestall the obvious response that the CEGB's hands are tied by the technical requirements of the nuclear siting process to such an extent that between 90 and 99.99 per cent of all sites which may seem to be suitable on demographic grounds would be unacceptable, it is worth pointing out that in the US there are three absolute requirements: (1) demographic restrictions which are mainly concerned with limiting the possible consequences of potential accidents; (2) environmental restrictions; and (3) reactor cooling water availability requirements. All other potential constraints are of an economic nature. In the UK it would seem that these marginal or otherwise economic restrictions have been paramount but they have been implicit rather than explicit. However, it can be argued that the fine tuning of the economic performance of nuclear plant by using the locational decision to optimise engineering and transmission system characteristics is hardly relevant given the far larger historical uncertainties in the magnitude of construction delays, operational difficulties, and cost over-runs typically measured in units of 100 per cent! To this list could be added: fuel price fluctuations, unpredictable reprocessing charges, a largely unknown level of system reliability, the unknown costs of decommissioning and site restoration, as well as the costs of waste disposal and diseconomies resulting from forecasts of power flows 10 to 20 years ahead. It would seem that the locational decision is heavily biased towards CEGB convenience (either technical or operational) and against public safety in the form of seeking sites that minimise the likely impacts of accidents. It would appear also that the importance of what are in reality marginal economic factors have been exaggerated out of all proportion and given complete preference to other non-economic effects of locational decisions; such as populations at risk, degree of public opposition, and even environmental considerations.

If there is a high degree of uncertainty in many of the economic variables then why should so much attention be given to optimising variables which in the final analysis are not significant factors in the total plant cost equation? It would be better from the public safety point of view to forego some of the marginal savings by seeking sites which are safer and publicly more acceptable. One way of doing this is to broaden the site evaluation process to include explicit safety criteria consistent with the UK government's safety goals (HSE, 1979) – one of which states that all reasonably practicable steps should be taken to minimise the health consequences of any accident. Unfortunately, this has been interpreted as requiring the existence of an emergency plan, not the use of remoter sites. Reasonably 'remote' as a geographer might interpret it, does not necessarily mean the Outer Hebrides, but that sites could be ranked in terms of their likely health impacts should a 'standard' accident occur which could not be handled by the engineered safety measures. This argument is developed further in later chapters. An important first step in the re-education of those responsible for nuclear siting decisions is to demonstrate that even in this heavily populated island of ours there is a considerable degree of spatial freedom that the population geography of the country allows without going to stupid or absurd distal locations.

Searching for remote sites by computer

The only remaining task is to apply these computer search methods to look for locations that satisfy the remote demographic siting criteria that have been discussed. Their application to other siting criteria is described in later chapters. The general belief in the mid 1960s was that the remote siting criteria were so stringent that there were scarcely any sites at all which would not cause the most horrendously unacceptable environmental damage. It would seem that hardly anyone had any real ideas as to what 'unacceptable environmental damage' might consist of. Instead it has been very easy to claim that any damage was the best or least that could be achieved, and to hope that no one either noticed the cumulative effect of many seemingly small changes spread over a long period of time. So whilst the nuclear industry and successive governments have always claimed to be seeking to minimise environmental damage, in practice this was often used as a means of justifying the selection of a particular site, and because the selection had been so carefully made there was no need to evaluate any other sites. One imagines that had there been a more active concern for assessing the subjective impacts of nuclear developments upon the landscape there might well have been an early development of site evaluation methods and a greater emphasis on the comparative evaluation of alternative sites during public inquiries.

It is of considerable interest to perform computer searches using remote siting criteria as a means of identifying retrospectively where the feasible locations actually existed. For convenience the various remote siting criteria discussed earlier in this chapter can be categorised in the following ways: policy I is the Charlesworth and Gronow (1967) version of the Marley and Fry (1955) criteria, see Table 3.3; policy II is the version of Marley and Fry (1955) as described by Farmer and Fletcher (1959), see Table 3.5; policy III is the Farmer and Fletcher (1959) revised remote siting criteria, see Table 3.5; policy IV is the Farmer (1962) revised remote siting criteria; policy V is the same as policy I but with the 0.5 mile population limits removed, on the grounds that the values used here are arbitrary and the populations within this annulus are subject to large estimation errors; and policy VI is the same as IV but with the 0.5 mile limits removed. It is noted that the evaluation of these 6 remote siting policies was performed for 799,000 locations using fixed and rotated sectors. Table 3.9 shows the proportions of the coastal regions and the total land area that would be excluded by each of these policies. Figures 3.2 to 3.3 show the distribution of sites where nuclear power stations would be prohibited on strictly demographic grounds. Appendix 1 gives a breakdown of the areas excluded by county.

TABLE 3.9 *Areas of the UK excluded by various remote siting criteria*

policy variant	coastal sites sites	%	total land excluded area	%
I	4884	39.7	96751	42.4
II	4881	39.6	96527	42.3
III	4832	39.2	95329	41.7
IV	6042	49.1	126748	55.5
V	2386	19.3	46258	20.2
VI	2605	21.1	46436	20.3

The immediate impression from examining these distribution patterns is that the remote siting policies are fairly restrictive but less so if the 0.5 mile population limit is relaxed and Farmer (1962) revised criteria are used. However, it is clear that they are not as highly restrictive as thought and there is in fact no acute shortage of sites in coastal areas generally. The infeasible areas

119

FIGURE 3.2 Infeasible sites in Britain according to the remote siting criteria of Marley and Fry (policy V).

FIGURE 3.3 *Infeasible sites in Britain according to the revised remote siting criteria of Farmer (policy VI).*

121

are closely related to the population distribution of the UK, picking out the major concentrations.

Conclusions

The results of the computer analyses generally indicate that the restrictiveness of the remote siting criteria have been exaggerated; indeed when expressed on the same basis as the various other criteria in Chapter 6, the remote siting criteria are fairly relaxed. Additionally, the definition of remoteness is not sufficiently strict to offer any real factors of safety because of the limited distances used in the summations. It is perhaps just as well that the CEGB have never relied on siting as an important safety measure.

Appendix 1

TABLE 1 *Coastal areas of UK counties excluded by various remote siting criteria*

Region County	I	III	IV	VI
Scotland				
Borders	55	45	90	10
Central	100	100	97	75
Dumfries & Galloway	16	11	31	3
Fife	79	77	82	43
Grampian	51	45	65	17
Highland	8	6	18	1
Lothian	58	56	63	33
Strathclyde	16	13	26	7
Tayside	73	73	67	43
Islands	6	1	23	0
East Anglia				
Cambridgeshire	0	0	0	0
Norfolk	68	66	62	30
Suffolk	66	63	79	31
East Midlands				
Derbyshire	0	0	0	0
Leicestershire	0	0	0	0
Lincolnshire	22	20	27	0
Northamptonshire	0	0	0	0
Nottinghamshire	0	0	0	0
North				
Cleveland	100	100	82	72
Cumbria	56	54	71	25

TABLE 1 *cont.*

Region County	I	III	IV	VI
Durham	100	100	100	100
Northumberland	58	54	69	31
Tyne & Wear	100	100	100	100
North West				
Cheshire	100	100	94	90
Greater Manchester	0	0	0	0
Lancashire	90	90	88	72
Merseyside	98	98	100	95
South East				
Bedfordshire	0	0	0	0
Berkshire	0	0	0	0
Buckinghamshire	0	0	0	0
East Sussex	91	89	90	63
Essex	74	73	70	45
London	100	100	100	100
Hampshire	93	93	90	80
Hertfordshire	0	0	0	0
Isle of Wight	75	72	86	28
Kent	93	92	72	54
Oxfordshire	0	0	0	0
Surrey	0	0	0	0
West Sussex	98	97	100	69
South West				
Avon	98	98	88	72
Cornwall	63	57	77	19
Devon	71	67	87	40
Dorset	76	75	71	53
Gloucestershire	46	40	80	15
Somerset	66	58	84	28
Wiltshire	0	0	0	0
West Midlands				
Hereford & Worcester	0	0	0	0
Salop	0	0	0	0
Staffordshire	0	0	0	0
Warwickshire	0	0	0	0
West Midlands	0	0	0	0
Yorkshire & Humberside				
Humberside	66	64	70	38
North Yorkshire	73	68	77	30
South Yorkshire	0	0	0	0
West Yorkshire	0	0	0	0
Wales				
Clwyd	100	100	97	72
Dyfed	36	31	59	12

123

TABLE 1 *cont.*

Region County	I	III	IV	VI
Gwent	76	75	75	34
Gwynedd	48	44	66	13
Mid Glamorgan	81	81	81	36
Powys	0	0	0	0
South Glamorgan	94	94	98	81

TABLE 2 *Total areas of UK counties excluded by various remote siting policies*

Region County	I	III	IV	VI
Scotland				
Borders	12	10	19	4
Central	37	36	36	19
Dumfries & Galloway	72	69	71	37
Fife	72	69	71	37
Grampian	14	13	24	3
Highland	3	2	6	0
Lothian	42	41	46	28
Strathclyde	25	24	28	14
Tayside	20	19	25	7
Islands	4	2	12	0
East Anglia				
Cambridgeshire	59	53	79	20
Norfolk	50	39	81	10
Suffolk	58	48	88	14
East Midlands				
Derbyshire	72	68	82	38
Leicestershire	78	74	89	38
Lincolnshire	44	36	72	10
Northamptonshire	68	63	86	26
Nottinghamshire	82	78	90	51
North				
Cleveland	82	81	84	58
Cumbria	27	22	47	8
Durham	52	50	55	32
Northumberland	19	17	30	7
Tyne & Wear	100	100	100	97
North West				
Cheshire	79	74	92	45
Greater Manchester	100	100	0	97

TABLE 2 *cont.*

Region County	I	III	IV	VI
Lancashire	78	76	80	52
Merseyside	100	100	0	98
South East				
Bedfordshire	90	87	94	38
Berkshire	79	74	94	48
Buckinghamshire	83	79	93	42
East Sussex	80	74	92	36
Essex	82	77	93	46
London	100	100	100	0
Hampshire	70	64	84	35
Hertfordshire	94	93	98	70
Isle of Wight	72	62	89	31
Kent	82	78	92	37
Oxfordshire	71	66	89	28
Surrey	93	90	0	63
West Sussex	79	75	90	36
South West				
Avon	92	90	95	56
Cornwall	48	41	74	14
Devon	40	34	64	14
Dorset	52	44	78	21
Gloucestershire	60	53	86	21
Somerset	52	44	78	15
Wiltshire	55	46	78	22
West Midlands				
Hereford & Worcester	47	38	80	16
Salop	37	29	71	12
Staffordshire	81	77	92	48
Warwickshire	77	72	92	33
West Midlands	97	96	0	80
Yorkshire & Humberside				
Humberside	57	52	71	19
North Yorkshire	34	28	55	10
South Yorkshire	87	86	84	68
West Yorkshire	91	91	89	76
Wales				
Clwyd	33	28	58	16
Dyfed	24	18	51	6
Gwent	66	63	82	46
Gwynedd	26	21	46	3
Mid Glamorgan	97	97	93	77
Powys	13	11	26	3
South Glamorgan	87	85	92	60

4 Relaxed siting policies in the UK

By the late 1950s a number of problems had started to emerge. It was now recognised that there was no urgent need for nuclear power and that it was a high cost technology. A Select Committee Report (1967) argued that the dominant motive behind the first nuclear power programme was as assumed physical shortage of coal and oil and that cost was a secondary matter. Yet in 1960 the CEGB, who clearly believed in the early cost estimates, secured a downward revision of the 1957 plan (Burns, 1967, p. 91). There had been cost and time over-runs on the MAGNOX stations and the first plants to be finished had operating costs of 62 to 82 per cent greater than the 1957 forecasts. Additionally, the CEGB seemed to disagree with the UKAEA regarding the choice of the next reactor system. In 1962 the CEGB preferred the CANDU design, by 1964 the BWR was highly favoured. Meanwhile the UKAEA preferred their own creation, the AGR. Despite growing doubts about the need for nuclear power, the second nuclear power programme was formally announced in a 1964 White Paper (Cmnd 2335). Some 5000MW(e), later increased to 8000MW(e), would be ordered to begin operation in the period 1970-75. The choice of reactor type, either LWR (i.e. PWR or BWR) or AGR would be decided later. There was no urgent need for additional power generation that could be justified in terms of demand (until this period demand had been growing almost exponentially) and the White Paper suggested that an important justification was the need to give orders to engineering firms to keep a nuclear capacity available for the future. There was also a hope of securing export markets (Select Committee on Science and Technology, 1969).

The CEGB had to evaluate both an AGR and a BWR design before the final decision as to reactor type would be made.

Perhaps surprisingly the CEGB preferred the AGR design which was regarded as providing cheaper electricity, having the greater potential for further development, and having no technical problems that could not be solved readily (CEGB, 1965). However, it seems that the comparison was most unfair. The AGR design existed only in an incomplete form and the comparison was based on extrapolations to a 600MW station from a 30MW prototype, and assumed a number of possible new developments. The report seemed to be biased and based on incorrect assumptions (Burns, 1967). It seems that the choice of the AGR was only made under the strongest pressure from government. In the UK it is the government and not the CEGB who select reactor types; government is advised by the UKAEA and enforce their decision on the CEGB through the latter's need to receive government sanctions of capital investments. So on the basis of the CEGB's (1965) report the second nuclear power programme was based on the AGR with a number of orders being made over the period 1965-70.

The result has been termed the 'great AGR disaster' (Burns, 1978). The first AGR station at Dungeness was ordered without any commercial scale prototype, without any experimental verification of various aspects of the proposed reactor, at a time when there was little experience of concrete pressure vessels, and construction started without a complete design. The disaster was compounded by the subsequent decision to order four more plants – Hunterston, Hinkley Point, Hartlepool, and Heysham – based on two different designs, shortly after the first but before there was any feedback. There were also no longer any indications that there might be an export market. The orders continued despite knowledge that the initial Dungeness design had been extensively changed and that the AGR was now looking like a high cost option compared with the SGHWR; the latter was being developed by the UKAEA as a reserve against the improbable failure of the AGR.

Seemingly the CEGB had a free choice but it was not clear as to how free the CEGB really were to make important reactor choice decisions. They were certainly treated as subordinate participants in the early discussions of atomic energy policy. The first 10-year programme of 1955 was shown to its predecessor only for comment and the CEGB expressed a preference for a smaller programme than that of 1957. In 1960 they had at least managed to secure a downward revision in the size of the programme (Burns, 1967). The 1965 decision was apparently only made under the strongest pressure from the UKAEA and in the event the CEGB were probably right. Table 4.1 shows the order

TABLE 4.1 *Order and operation dates for AGR reactors*

Site	Date of order	date operational
Dungeness B	1965	1984
Hartlepool	1969	1984
Heysham A	1970	1984
Heysham B	1978	1987
Hinkley Point B	1966	1976
Hunterston B	1967	1976
Torness A	1978	1987

dates and the current status of the AGR reactor programme of 1964.

By 1969 it seems that the question of reactor choice was again under review. The AGR programme was suffering from various problems due to delays, cost overruns, and design changes. By now the CEGB preferred the SGHWR and the HTR (High Temperature Gas-cooled Reactor). In 1970 they invited tenders for a HTR station at Oldbury, but in 1971 design problems resulted in its postponement. The Scotland Hydro-Electric Board planned a SGHWR at Stake Ness but this was cancelled in 1972 due to the lack of government support. Meanwhile the CEGB deferred the 1970 Sizewell AGR on the grounds that there was no longer any need for additional capacity and seemingly prefered the PWR. It announced in 1973 that it would order no more AGRs until one worked! Another clash of interests now arose because the UKAEA wanted the SGHWR to be the next recommended reactor type, as a stop-gap measure whilst the development of the FBR continued.

In 1971 the CEGB saw need for between 1 and 4 new stations by 1980. However, the oil crisis led to suggestions in 1973 for a 45GW programme by 1982, initially based on 18 PWRs and one HTR. The SSEB were believed to be planning another 8 stations, so there was a possibility of 26 new nuclear stations with as many as 52PWRs. The justification was economic and the effects of the oil crisis. However, at that time the safety of the PWR was seen as an unresolved issue, at the same time it was thought that a programme based on the SGHWR would lead to export orders in the Third World.

In 1974 the government announced the third nuclear power programme involving the construction of a 4GW of SGHWRs. Once again the utilities (the CEGB and the SSEB) were to be forced to buy a reactor they did not want, once it had been declared 'safe' by the Nuclear Installation Inspectorate (NII). Two stations were approved: a 4-reactor 660MW(e) SGHWR station at Sizewell and a 2-reactor station at Torness. One problem was that the SGHWR still required further development, and a commercial scale prototype could not be ordered quickly. By 1976 there had been some technical slippage and the projected costs had risen alarmingly; partly because of the need to incorporate additional safety measures and partly because of design modifications to meet CEGB needs. These problems coincided with a UKAEA report that the PWR safety problems could be overcome (Marshall, 1976) and with advice that the SGHWR should be dropped in favour of the PWR (Select Committee, 1976). Finally, in 1978, the government cancelled the SGHWR programme and gave formal approval for two more AGRs (Heysham B and Torness). The pro-PWR lobby had exerted immense pressure but this had been resisted to some degree. The PWR was still not viewed as a viable option because of the need for design changes to meet UK safety standards. However, approval was given for a PWR order if the necessary design work could be completed satisfactorily and if safety approval was obtained, subject to a later decision by the government of the day.

In March 1979 the Three Mile Island incident in the USA was followed a few months later by a UK government announcement that, subject to safety clearance and a public inquiry, the PWR would be the next reactor ordered. Furthermore, as an indication for planning purposes, the government announced an intention (but no commitment) to consider a programme of 15GW of PWRs over the 10 years 1982-92, with perhaps another 10GW by 2000 AD. Since then there has been considerable slippage. The economic recession and the de-industrialisation of several parts of the UK have combined to reduce the demand for electricity. Additionally, the public inquiry at Sizewell into the proposal to build the first PWR looks like lasting until 1985; while the NII have not yet given approval to the proposed British PWR. At the same time the prospects for a commercial FBR programme have diminished somewhat although it is still envisaged that one will be forthcoming in the 1990s.

Some questions still remain about PWR safety and there are some doubts about the ability of the UK to build a large number of PWR reactors simultaneously (Taylor, 1980). It also seems that the safety advantages of the gas-cooled reactor are being rediscovered, particularly in the US. It is more than a little ironic that in the US the PWR boom ceased in the mid 1970s while in the UK it is about to start. The CEGB are firmly committed to nuclear electricity and a fourth nuclear power programme is needed to provide enough plutonium for the initial fuel charges for the long dreamt-of programme of commercial fast breeder reactors.

It is very reassuring to note that whilst all the debate about reactor choice was occurring and there were immense problems in finishing the AGR programme, there was never any doubt that public safety was being preserved. It is true that there is an absence of published safety studies for all the existing UK reactor types, and that most of the designs were one-off prototypes. Nevertheless, all the commercial reactors are completely safe because the operators and the government safety agencies say they are. It is largely a matter of faith. An indication of this can be seen in the minutes of Durham County Council Planning Committee when they were considering whether to object to the proposed Hartlepool station in 1968. The CEGB and government representatives were able to give assurances that

. . . the latest design of building, plant, and equipment were such that there was no possibility of harmful effects being brought upon local people (minute dated 17 April, 1968).

The fact that the first full-scale AGR reactor did not start operations until almost 10 years later is hardly a relevant factor. Note also that 'no possibility' of harm rather than 'a very low probability of harm' was used to convey the nature of the risks to the local politicians. Yet Hartlepool is probably the most heavily built up site in the UK. It also has the distinction of having the proposed 'new style' pressure vessel being declared 'unsafe' by the NII. As ever, the safety aspects of nuclear power are given the lowest possible degree of public exposure. The risks, which are probably still not fully understood, are totally discounted to avoid an adverse public reaction.

Table 4.2 summarises the various nuclear power programmes. As Figure 4.1 shows the existing nuclear stations are generally either in areas with no coal supplies; most of the MAGNOX stations and the first three AGRs; or are located near to large centres of demand (Hartlepool). The early stations were virtually restricted to coastal locations because of the need for large quantities of cooling water. The AGRs need far less water than the MAGNOX reactors and it seems likely that in the future increasing use will be made of existing non-nuclear sites for new nuclear developments. The only new nuclear greenfield site to be announced is at Druridge Bay; a coalfield location. As Catchpole and Jenkins (1977) put it:

TABLE 4.2 *UK Nuclear Power Programmes*

Programme	Year	Size in MWe
1	1955-65	1,500 to 2,000
	1957-65	5,000 to 6,000
	1957-66	5,000 to 6,000
	1960-68	5,000
2	1964-75	5,000
	1965-75	8,000
3	1974-78	4,000
	1978	3,200
4	1979-91	15,000
	1983-93?	10,000
	1986-99?	4,000 to 8,000

UK strategy for siting plants which featured in the first and second nuclear programmes was governed principally by the geographical pattern of fossil fuel sources in relation to the electrical load centres, safety and amenity considerations and cooling water availability (p. 180).

The third nuclear programme largely dropped the safety requirements because the AGR design was declared suitable for urban and semi-urban sites. The fourth nuclear programme, when it appears, will probably emphasise load centres and amenity considerations rather than public safety and public acceptability aspects. This is likely to prove to be a major error. If 'good' compromise sites cannot be found (and this is by no means certain), then there should be a concentration on increasing the remoteness of reactor siting even if it is at the expense of amenity and proximity to possible load centres.

Announcement of a relaxed siting policy

A statement by the Minister of Power on 6 February, 1968 to the House of Commons is usually regarded as marking the beginning of a new era of relaxed siting policy for nuclear power stations. The Minister told the House:

> . . . gas-cooled nuclear reactors in prestressed concrete pressure vessels could be constructed and operated much nearer to built-up areas than so far permitted (Hansard, 6 February, 1968).

131

FIGURE 4.1 Locations of current nuclear reactors in the UK

Hartlepool and Heysham were quoted as examples of acceptable sites. Although it made possible the siting of AGRs at these places, these sites had in fact been proposed for AGR plant long before the Minister's announcement; indeed applications for Section 3 consent under the 1909 Electricity Lighting Act and for nuclear site licences had been made on 7 Februasry, 1967 for Heysham and 8 March, 1967 for Hartlepool (Adams and Faux, 1969, p. 38). The Minister's statement was in fact the conclusion of a long internal debate in the period 1965-67 about possible revisions to the Farmer (1962) 'remote' siting criteria in the light of recent advances in reactor safety. The revised policy removed the conditions that would otherwise have prohibited the issue of licences for Hartlepool and Heysham. These sites are particularly significant because they were the first attempts at metropolitan siting in the UK; this was needed to keep pace with similar applications being made in the US (which were in fact all rejected or cancelled later) and because the Hartlepool site was a coalfield location. Nuclear power it seemed had become competitive with coal fired plant in coalfield locations.

The Minister did not specify any particular siting criteria or that the Hartlepool and Heysham sites were limiting cases. In fact the decision had been made by the full cabinet prior to the Minister's appointment. In a later statement made on 12 March, 1968 it was stated that '. . . it was not yet contemplated that stations could be licensed within a mile or two of full urban density' (Hansard, 12 March, 1968). Although no definitions were given of site requirements, the interpretation then made was that the existing emergency measures, comprising stable iodine distribution and public evacuation, must be practicable within 2 hours for any 30 degree sector up to two-thirds of a mile from the reactor, and that no special difficulties in the police control of any emergency should be presented by either the topography or population characteristics up to a distance of two miles. It should be noted that the question of the range over which the emergency measures are planned in advance is related to what is considered practicable and is not related to the characteristics of the plant (Fryer, 1969).

The parliamentary announcements made it clear that no figure had been prescribed as a maximum population density and that applications for site licences would be considered individually. If a prospective licensee believes that the criteria regarding emergency arrangements are satisfied by a site then the next step is empirical. He finds out whether the site will receive a licence by applying for one. In practice preliminary discussions between the CEGB and the Nuclear Installations Inspectorate would be held to try and avoid public embarrassment. The Inspectorate of Nuclear Installations has a powerful control on siting decisions. Hartlepool and Heysham are again of interest as the first sites where the provisions of the 1960 Nuclear Installations Act was applied in full prior to the site being adopted. In theory, the 1960 Act required that sites receive a nuclear licence prior to any decisions to construct a reactor there. This approval was in fact given to the MAGNOX stations at Berkeley, Bradwell, Trawsfynydd, and Hunterston after they had been built! At Sizewell, Oldbury, and Wylfa, although construction did not begin until after site licences were issued, nevertheless the siting decision had already been accepted by government prior to that (Cave and Halliday, 1969).

The suitability of the first two relaxed sites, and so far the only ones, seems to have been a matter of ministerial fiat. They were simply declared to be suitable sites presumably on the basis of various unpublished submissions to government. It was rumoured that in fact the Nuclear Installations Inspectorate opposed the sites but that the pro-nuclear industrial lobby had won the behind-the-scenes political arguments. The decision was supported by a belief that the AGR economics were competitive even on a coalfield location, that they were opportunities for exports to Japan and Austria, and that siting was no longer a relevant safety measure following the recent change in emphasis towards probabilistic safety assessments by the UKAEA (Farmer, 1967).

A case for unrestricted siting

In fact by the late 1960s it almost seemed that all siting restrictions would be removed. This was implicit in the 1968 announcements but there was a subsequent retreat back to formal criteria. There was certainly strong industry pressure for unrestricted siting. Adams and Faux (1969) claimed that a 1963 relaxation which allowed MAGNOX reactors with concrete pressure vessels to be located on Farmer class II sites did not widen the range of choice sufficient for the second nuclear power programme. The use of prestressed concrete pressure vessels for the later MAGNOX stations was apparently an innovation introduced by the builders who decided to ignore the CEGB specifications in order to build larger reactors with higher gas pressures than could be obtained with steel pressure vessels (Vaughan and Joss, 1964). As Sir Owen Saunders, chairman of the Nuclear Safety Advisory Committee explained:

To continue (the remote siting policy) leads to difficulties, not only because of the running out of suitable sites, but also because of the adverse effect on amenities in the countryside and, moreover, the economic considerations must increasingly point to bringing the stations closer to the industrial centres which consume their product, always assuming that safety can be taken care of.

In the mid 1960s the prevailing engineering view was that safety was now fully taken care of by the AGR design, and that site isolation had only a small contribution to make. Certainly it was considered by the CEGB that prior to the Hartlepool and Heysham decisions, the existing siting criteria would have made integration of the new stations into the 400kV network progressively more difficult. Long transmission connections were seen as complicating matters and increasing costs, especially if amenity considerations were to result in the large scale use of underground cables.

The twin arguments, related to the shortage of satisfactory remote sites and the need to preserve amenity, eventually had the desired effect. Gronow 1969 wrote:

The very limited improvement in safety which could be achieved by chosing remote reactor sites could be considered against the social, economic and amenity advantages that may arise from the use of sites rather closer to urban areas (p. 551).

Why the Nuclear Installations Inspectorate (Gronow was an inspector) should be so concerned with amenity and economic considerations is not clear. Nevertheless, this is clearly an attempt to rationalise what really amounted to a reduction in public safety standards due to the erosion of the protection offered by distance and low population density. According to Adams and Faux (1969), the CEGB wanted a siting policy based on two concepts: an exclusion zone which would be located within the station fence and an emergency action area. It seemed that the Nuclear Installations Inspectorate were not willing to agree to this de-restricted siting policy.

There were also other reasons for wishing to be able to locate power reactors with no or minimal restrictions. They were discussed in the mid 1960s and no doubt the same points will again be raised in the not-too-distant future. Two of the most important are: (1) the use of reactors for purposes additional to simple electricity generation where it is expensive to site reactors far from load centres – for example, in district heating or the production of process steam for industrial applications; (2) for supplying electricity to extremely large load centres in situations where overhead transmission lines cannot be erected either for reasons of effect on amenities, or because of difficulties in obtaining the necessary wayleaves. The capital cost of underground cables is regarded as prohibitively expensive for large-scale use.

Other reasons might well be: (3) it is useful for public relations purposes if public opinion can be persuaded to accept nuclear reactors as just another form of industrial activity and this precludes the use of isolated sites which imply dangerous activities. (4) The CEGB own many sites on which they may wish to develop nuclear power as a replacement for obsolete conventional plant, especially in urban locations; this would also allow best use to be made of the existing transmission infrastructure. (5) It can be argued that in densely populated countries such as Britain nuclear power must attain a level of safety that will permit complete freedom in siting apart from a small exclusion zone focused on the reactor. This follows from the observation that the safety factors conferred by siting are far smaller than those obtainable by engineered safety measures and that in a densely populated country major accidents cannot be allowed to happen (Schwarzer, 1968). (6) Unrestricted siting would allow the best sites from an engineering viewpoint to be utilised. (7) Manufacturers claim that exports are dependent on the development of reactors which could be sited anywhere. Finally, (8) the development of the AGR with a high degree of engineered safety meant that the additional hazards resulting from unrestricted siting were small and could be offset by engineering design changes (Cave and Halliday, 1969).

It has been estimated that the total risks from nuclear power would be increased by only a factor of 10 for unrestricted siting (Schwarzer, 1968) or between 10 and 100 (Beattie, 1969), and that it should be possible to achieve a corresponding decrease in the probability of severe accidents by changes in the design which would not be unduly uneconomic. An additional argument might well be that if accident frequencies are already vanishingly small, say 10^{-8}, then even a two orders of magnitude increase would not mean very much. On the other hand the main problem with urban siting is that the effects of any accident would be more noticeable. It would be necessary for reactor systems to be devised that would give a sufficiently long warning period before any major release so that the necessary emergency measures could be implemented. The problems of large-scale and long-term evacuations of urban areas do not seem to have been seriously considered in the 1960s. Indeed it was thought that evacuations of people from within 3 to 4 miles at the most would eliminate the risk of radiation sickness from an accident on an urban site. One imagines that this is merely a reflection of the restricted nature of the design basis accident that was hypothesised.

In the UK it was the AGR that was being proposed as suitable for urban siting. The reasons were due to its inherent safety features, particularly: the use of an integral pressure circuit arrangement with a concrete pressure vessel that greatly reduces the risk of catastrophic failure; the use of the moderator as a heat sink and the use of fuel and cladding that can withstand melting due to decay heat for at least 4 hours after shutdown; the retention of the gaseous and volatile fission products by the fuel; the fact that there is no risk of large sudden temperature changes; and that there is sufficient natural circulation to remove decay heat. The result was that the most severe form of maximum credible accident was thought unlikely to result in the release of sufficient fission products to cause any need for emergency evacuation. On the other hand the AGR has a positive temperature coefficient and there is a possibility of an exothermic reaction between graphite and air should any air enter the reactor. Concern has also been expressed about the possibility of the gas baffle hindering the operation of the main shut-down system (PERG, 1980).

Prudence suggests that the era of unrestricted siting has gone for ever, but it may well return should a reactor system be developed that is completely safe. On the other hand the trend towards multiple developments and probably towards fast breeder reactors are factors which favour more isolated sites for security purposes as well as safety. However, had there been a large UK reactor programme in the early 1970s, it would probably have involved a large number of metropolitan sites being developed.

Towards a formal relaxed siting policy

The relaxed siting policy that emerged in 1968 was a result of advice received from the Nuclear Safety Advisory Committee which was appointed by government for advice on nuclear safety. The review, which was never published, reaffirmed the principle that the major contribution to public safety lay in the standards achieved in the design, construction, and operation of reactors. Doubtless government was also advised by the UKAEA that the AGR reactor they had designed was safe and that consequently siting restrictions were no longer relevant. Farmer (1967), head of the Health and Safety Branch of the UKAEA, explained 'It is no use gaining a factor of 3 or 5 in safety by siting and losing a factor of 10 or 100 through lack of attention in reactor engineering' (p. 326). It is not that anyone has ever suggested that siting should be the only safety measure but that it is possible to gain a factor of 10 or 100 by better engineering compared with 2 or 3 by siting measures. The government of the day, advised by the experts, clearly agreed.

The initial specification of the relaxed siting policy was in terms of emergency planning constraints and a similarity with the semi-urban sites at Hartlepool and Heysham. One suspects that the evacuation constraint was a cosmetic sop to public confidence that was necessary to provide a degree of continuity with previous sites. There was in fact no known design basis accident which could lead to environmental contamination, and thus the need for public evacuation, from an AGR reactor (Macdonald et al., 1977). The most severe accident at a MAGNOX station was thought to require evacuation from 1 to 2 mile distance from the reactor. The two-thirds of a mile quick evacuation limit specified for AGRs is not particularly relevant since few people, if any, other than utility employees, live that close to nuclear power stations. The site itself provides a zero population exclusion zone. The ability of the police to extend emergency measures up to 2 miles is also a very weak and vague constraint. It merely implies the absence of low mobility populations; even then, the police may still not be able to cope.

The emergency planning restrictions would have become more binding if inner city or redevelopment sites had been sought, but there is no evidence to indicate that such constraints would preclude the development of urban sites if government wanted to use such sites for nuclear power.

The 1968 announcement also indicated that sites with similar population characteristics to Hartlepool and Heysham would be considered suitable. The problem was how to assess any given site for these characteristics. Innsworke Point was one such site. The CEGB had an outstanding consent for an oil fired power station but the site was regarded as being too densely populated for an AGR. So it seems that a difference of interpretation existed between the Nuclear Installations Inspectorate who viewed Hartlepool and Heysham as limiting cases, and the CEGB who would have preferred a combination of an exclusion zone and emergency planning restrictions. There was clearly a need for some numerical siting criteria which would specify acceptable population distributions, both to guide prospective licensees in the examination of potential sites and to set standards against which applications might be judged. Criteria were also needed to provide a basis for controlling new population developments around existing sites. It was, as Gronow and Gausden (1973) wrote, '. . . essential to provide clear guidance on the development restrictions associated with a nuclear plant' (p. 532). Once a site was accepted the same criteria that were used to establish its acceptability could be used to identify the degree of population growth that would be acceptable. These criteria seemed to emerge after the 1968 announcement and were certainly in use by 1973; indeed, the method used had first been published in 1967.

At the Connagh's Quay public inquiry the statement by the Department of Trade and Industry (then responsible for nuclear safety) stated the government's policy on siting nuclear power stations. It was written:

> Each proposed site must be considered on its particular merits but the sites at Hartlepool and Heysham were given as examples of acceptable sites. Population density close to a site had to be sufficiently light to enable effective emergency countermeasures such as the evacuation of people from the area in the very unlikely event of an accidental release of radioactivity having an effect beyond the station boundary.

Furthermore,

> A site is only acceptable if the surrounding population together with all likely future developments will remain consistent with this siting policy over the life of the station. For this purpose a proposed site is assessed by comparing the expected future total population distribution around it with established criteria by a standardising method which gives greater relative importance to those people closer to the site than those further away. . . . The distribution of population around a site is an important factor in the assessment and it is therefore not possible to give precise limits on population numbers unless the location and the density are also clearly defined (Department of Trade and Industry, 1972, p. 2).

However, population numbers and distribution are not the only important factors and '. . . it would not necessarily follow because a site fell within the acceptable population distribution then it would always be acceptable' (p. 2). Examples were given of particular topographies, local communications, population mobility and any other special features which would have to be evaluated for their probable effect on any emergency measures close to the station. In addition, 'The Government siting policy

139

aims at approving only those sites which would not restrict probable development around them' (p. 2). To ensure that future development in the immediate vicinity of a nuclear station does not prejudice the effectiveness of the emergency counter-measures local authorities are requested to consult the relevant government departments on certain categories of planning application within a radius of 2 miles.

The standardising method referred to here was devised by Charlesworth and Gronow (1967); see also Gronow (1969) and Gronow and Gausden (1973). It is still in use today (HSE, 1982a). The method is a variation of Farmer's (1962) site rating index. It was modified to take into account both the population within a 30 degree sector and all-round population totals. This is necessary because near urban areas it is unlikely that the bulk of the population would be concentrated in any single 30 degree sector. Indeed it is a very characteristic feature of semi-urban sites that they can have population centres distributed around them in all directions and would, therefore, be more likely to have persons affected by a release of radiation than a remote site with only one nearby population centre. The assumption made about the amount of radiation to be released was also modified. Farmer (1962) used a 1 curie release under Pasquill category F weather. Charlesworth and Gronow (1967) replaced this by a 1000 curie release also under Pasquill category F weather conditions.

It was considered that siting measures would be quite ineffective for large radiation releases, irrespective of location, large numbers of people would be affected in a densely populated country. Large releases must, therefore, be prevented from occurring by engineering design measures. On the other hand, minor largely unforeseen accidents involving irradiated fuel cannot be discounted. Such accidents might release a few thousand curies of gaseous and volatile fission products (Charlesworth and Gronow, 1967, p. 148). For these small accidents siting does make a difference and can mitigate the resulting health consequences. Even for small releases, quite large numbers of people might be exposed to low levels of radiation and long-term cancer risks.

It was considered that the most extensive hazard to the public is caused by the inhalation of the radioactive isotope iodine 131. Methods exist for estimating the cumulative thyroid dose from the concentration of airborne material that would be received by a 'standard person'; allowance being made for the variation in sensitivity by different age groups. The release is assumed to take place over a few hours and the lateral speed of the plume is restricted to 30 degrees. The results can be expressed as cumulative population doses to the thyroid and plotted as functions of distance for both the most densely populated 30 degree sector (sector risk factors) and summed for various distance bands (site risk factors). The resulting sets of weighting factors are shown in Table 4.3 for the 7 distance bands that are used.

TABLE 4.3 *Site and sector weighting factors for the relaxed siting criteria*

Distance band (miles)	Sector weights	Sector weights
0-1	40.0	32.4
1-1.5	22.5	17.6
1.5-2	13.75	10.5
2-3	8.75	6.4
3-5	4.75	3.3
5-10	2.0	1.31
10-20	0.75	0.46

The sector characteristics of a proposed site are obtained by summing the products of the populations within each sector and distance band and multiplying by the associated weighting factors. The site factor (u) for any site is given by

$$u = \sum_{i}^{7} W_i \sum_{j}^{12} P_{ij} / 1000000$$

where
P_{ij} is the population located within the ith distance band and the jth 30 degree sector; and
W_i is the site weighting factor for the ith distance band.
The sector factor (v) involves

$$v = \max_{i} [\sum_{i}^{7} P_{ij} W^*_i / 1000000] \text{ for } j = 1,2,...,12$$

where
W^*_i is the sector weighting factor for the ith distance band.
The summation of the 7 distance bands is based on Charlesworth and Gronow (1967). It is interesting that the Department of

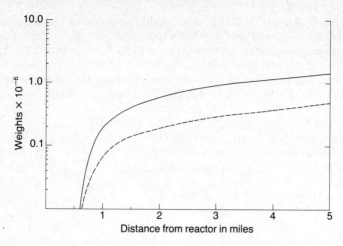

FIGURE 4.2 *Site and sector limiting curves for AGR reactors*

Trade and Industry (1972) evidence at the Connagh's Quay Inquiry mentions distance bands covering 0 to 5 miles only.

The resulting site and sector characteristic curves can be plotted and compared with two limiting distributions; see Figure 4.2. These limiting curves can also be expressed as point values for each of the 7 distance bands; see Table 4.4.

TABLE 4.4 *Limiting site and sector values*

Distance band (miles)	Sector limit	Site limit
0-1	.063	.190
1-1.5	.137	.411
1.5-2	.198	.595
2-3	.305	.917
3-5	.482	1.448
5-10	.812	2.436
10-20	1.274	3.823

These limiting values supposedly represent the risk that government is prepared to accept for a specific reactor type. How these values were obtained is not known, but it is known that the curves were calibrated to ensure that all current sites, as well as the Hartlepool and Heysham sites, would not violate the limiting values. Charlesworth and Gronow (1967) report a large number of characteristic curves for existing, proposed, and hypothetical sites. Their analysis indicated that the risk factors varied by over three orders of magnitude. Gronow and Gausden (1973) commented that for practical purposes sites falling within one order of magnitude were considered to be of comparable intrinsic risk. In this way three broad categories of site could be identified. The Farmer class I sites chosen for the first reactors had the lowest ranges of risk. The class III metropolitan or near metropolitan sites had the highest, and in between were the class II sites including Hartlepool and Heysham. All this was highly subjective but it provides a neat way of ensuring continuity with the earlier Farmer (1962) criteria.

The other subjective decision of any importance concerns the choice of a 1,000 curie release. The weights reflect this assumption. It is more than a coincidence that Farmer (1967) also used a 1,000 curie release to demonstrate his new safety principle. According to his proposed reactor release criterion, a release of 1,000 curies of iodine 131 would have a 1 in 3 chance of occurring in 1,000 years of reactor operation; a state which would be reached world-wide by 2000 AD. A release of 1,000 curies was also the smallest accident large enough in Farmer's terminology to be considered more than a nuisance. Farmer's method (1967) provides a neat approach whereby site population characteristics and reactor design can be matched in an integrated fashion; but it is, of course, highly subjective.

Gronow and Gausden (1973) describe the application of these relaxed siting criteria as follows:

> The suitability of any proposed site is first tested by comparing its characteristic curves, derived from an analysis of the density and distribution of the surrounding population including projected development, with the limiting characteristics. The derived characteristics of the proposed site should not exceed either the limiting site or sector characteristics (p. 532).

In practice it seems that interpretation is not quite so straightforward.

Problems of applying and interpreting the relaxed siting criteria

One major problem involves the identification of the most densely populated 30 degree sector. There are two difficulties. First, there is the problem of using an arbitrary origin for the 30 degree sectors. Another concerns the choice of the 30 degree

143

sector which is the most densely populated. This is often difficult since the sectors with the highest close-in populations will sometimes not be the most densely populated out to the 20-mile distance limit. This is due to the typically discontinuous patterns of urban development found on peri-urban sites. Other problems relate to the interpretation of the limiting curves.

The critical words are 'should not exceed' the site or sector limiting values. This would imply that any points on the characteristic curve that exceed the limiting values would render the site unacceptable. However, another interpretation is that the occasional spike above the limits does not matter provided the cumulative site and sector risk values are within the limits at the 20 mile boundary. This is understood to be how the criteria have been interpreted in practice, but with the caveat that any spikes above the limits close to – say within 5 miles – of a reactor would be closely looked at in case it made implementation of the emergency plan difficult. These differences in application are quite fundamental. The latter implies that it is the total 'risk' that is important and that an excess of risk in one distance annulus can be offset against a deficiency in another. The importance of these differences can be illustrated by converting the limiting values in Table 4.4 into equivalent population levels; see Shaw and Palabrica (1974). Two sets of population limits are obtained; one is the maximum populations allowed in each distance band if all closer distance bands have populations set at their maximum levels, and the second is the maximum population allowed in each distance band if the populations in all preceding distance bands are zero; see Table 4.5.

The differences between the population levels for the two assumptions reflect the presence or absence of compensatory effects which are implicit in the interpretation of the criteria. For example, column 4 in Table 4.5 gives a population of 5.097 million as the maximum population allowable in the 10 to 20 mile distance band if the population within the 0 to 10 mile annulus is zero. This could be applicable to an offshore or floating reactor, both of which have been proposed in the US. In principle the criteria would allow such a development to be located 10 miles from even the largest population concentrations; for example, London or Edinburgh. The values given in column 2 of Table 4.4 are the maximum population that would be allowed, for example, in the 10 to 20 mile annulus if all preceding distance bands were set at their maximum levels; if some or all were less, then these levels could be increased, reaching in the limiting case the values shown in column 4. So the relaxed siting criteria currently used in the UK are in fact far more complex than they may appear at first

TABLE 4.5 *Some radial and sector population limits for the relaxed siting criteria*

Distance band	Assuming uniform maximum populations in all distance bands		Maximum population allowed if all preceding distance bands have zero populations	
(Miles)	Sector	Site	Sector	Site
0-1	1,944	4,750	1,944	4,750
1-1.5	4,205	9,822	7,784	18,266
1.5-2	5,810	13,382	18,857	43,272
2-3	16,718	36,800	47,656	104,800
3-5	53,636	111,789	146,060	304,842
5-10	251,908	494,000	619,847	1,218,000
10-20	1,004,348	1,849,333	2,769,565	5,097,333
Total at 20 miles	1,338,569	2,519,876	3,611,713	6,791,263

sight. The critical values are site-dependent, they reflect the density and the distribution of population, and the manner in which they are applied.

The final uncertainty concerns the distance over which the site and sector risk calculations are performed. The Department of Trade and Industry referred to previously only show limits extending to 5 miles. It may be that this is the principal area of concern. These uncertainties about the manner by which the criteria are applied exist because the Nuclear Installations Inspectorate refuses to discuss anything related to their siting criteria. It may be deliberate in order to allow maximum flexibility in assessing the merits of various proposed sites. Some justification for this flexible interpretation is given by Charlesworth and Gronow (1967) when they distinguish between close-in population characteristics and the 20 mile characteristics. Indeed they consider that there are three possible orderings of sites according to demographic merit; in terms of the population as a whole, the most densely populated sector, and the exposure of individuals living close to the site. The problem as they see it, is that

. . . there are too few practical sites high in the merit order on all three counts to form an adequate basis for the substantial development programme which is envisaged for nuclear energy in the UK (p. 161).

145

The siting criteria they devised certainly seem to give sufficient flexibility to cover both semi-urban and urban sites. The latter would be possible, subject to any restrictions necessary to ensure an operational emergency plan, in or near many major urban areas by the simple expedient of adopting the most liberal or relaxed interpretation of the siting criteria. It is not a deliberate policy to exploit the criteria in this way, yet the flexibility is there should the need arise. The critical limiting values and the weighting factors could also be relaxed further by updating the atmospheric diffusion model used by Charlesworth and Gronow in 1967. There have been major changes in computer technology and in the mathematical sophistication of diffusion models since then. Were they to be incorporated into the relaxed siting criteria then even larger population values would be permitted; see, for example, Shaw and Palabrica (1974). However, such refinement would imply that someone, somewhere, actually believes that the hypothetical radiation release on which the criteria are based, is sensible and a meaningful basis for a reactor siting policy that seeks to control public exposure and risks. It is most unlikely that a complete reassessment of public risks would result in anything like the current criteria. There would be a major conflict between those who oppose any criteria not related to emergency measures and those who would wish to take into account not small accidents but major accidents which current reactor designs could not handle. There is further discussion of these points in a subsequent chapter.

Using the relaxed siting criteria to define remote sites

The CEGB seem to apply the relaxed siting criteria in the most stringent fashion. That is when they evaluate sites during the search phase they interpret the criteria as implying definite allowable population limits in both a cumulative fashion and in terms of the numbers allowed in each distance band. Furthermore, the siting guidelines they use internally in searches for sites for new thermal reactors are even more restrictive. The Nuclear Installations Inspectorate have a policy of siting the first few reactors of a type not previously built in the UK on 'remote' sites. This would normally mean one of the existing 'remote' sites, as defined using the 1955 criteria. Problems arise if there is need to find some new remote sites. The CEGB guidelines for new remote sites are based on the relaxed siting criteria but with the limits being reduced to one-third the normal values. Table 4.6 shows the population guidelines that seem to have been used by the CEGB since at least 1974.

TABLE 4.6 *CEGB populations guidelines for 'remote' sites*

Distance band (miles)	Population allowed in each distance Sector	Site	Cumulative populations allowed Sector	Site
0-1	648	1,575	648	1,575
1-1.5	1,420	3,288	2,068	4,863
1.5-2	1,904	4,436	3,972	9,299
2-3	5,625	12,228	9,597	21,527
3-5	17,878	37,263	27,475	58,790
5-10	83,969	165,000	111,444	223,790
10-20	334,782	616,000	446,226	839,790

Source: CEGB document 7.2.1974; but figures reworked to avoid rounding errors contained in original document.

It is clear that the suggested population limits offer a considerable degree of conservatism. This reflects the dilemma that accompanies any application for a site licence. The onus is on the applicant to avoid marginal sites which the Nuclear Installations Inspectorate may veto after expensive preliminary investigations have been carried out. The values shown in Table 4.6 may not be the upper limits for a new remote site, they merely provide a working definition of a 'remote' site that the CEGB think will eventually prove to be acceptable.

Some CEGB criticisms of the relaxed siting criteria

The search for new power station sites is a continuing responsibility within the CEGB. The need for extensive site investigations emphasises the need to ensure that work is not wasted on sites that are subsequently found to violate the siting criteria. Adams and Faux (1969) write

It is also important that sites should not be unnecessarily excluded from investigation because of the apparent inability to meet requirements put forward on the basis of safeguarding the public (p. 40).

Their complaint is basically that numerical siting criteria which are capable of being expressed in a rigorous form may seem attractive but in fact some effort is required to obtain the population data. It might be thought that this expense is compensated by their ability to yield an unambiguous answer as to whether a site has ratings above or below the permissible limits. In practice, they argue, this advantage is not so definite

147

because heed must be paid to the various borderline cases and to probable future developments. They conclude that

> The main disadvantage however is that a penalty is imposed on the supply system which cannot be shown to correspond with any clearcut gain in safety (p. 40).

The latter requires that the licensee looks at the more basic considerations underlying the criteria to find whether it is reasonable to advocate a site which fails to meet the standards being applied. In this way numerical siting criteria can be obstructive (if they are not binding) because the licensee could have saved time by considering the more basic questions, such as ease of evacuation, first. The solution according to Adams and Faux (1969) is to have no specific criteria other than a need for close-in evacuations. Such an approach is considered to be logically more defensible than the apparently more objective and systematic approaches which are based on population density statistics. They certainly have a point, especially if the numerical criteria are not going to be applied in a consistent and clearly defined fashion. If they are used as ad hoc guides then clearly other factors might be deemed more relevant to the CEGB and more significant to public safety.

The site rating approaches which were used by Farmer (1962) and are the basis for the current relaxed siting criteria can be subjected to a number of more general criticisms. They imply acceptance of two propositions: (1) that all reactors of a given class have the same maximum accident potential; and (2) that the site and sector factors provide a reasonable indication of possible levels of casualties albeit in a relative way. Adams and Faux (1969) maintain that both assumptions are questionable. In addition, (i) there are uncertainties in the calculation of casualties that are used in the site assessments; (ii) the low and imprecisely known probabilities of those accidents for which the calculation would be relevant make it very doubtful whether any large returns in public safety would follow from siting decisions made using these criteria. (iii) It can cause contradictory situations – for example a housing estate might be acceptable 3 km from a reactor in the most frequent wind direction, but not the same distance away in another direction because of the different population densities in the two sectors. (iv) The site rating numbers give an unjustified impression of accuracy in the assessment of risks when applied to the control of new developments. (v) Reactor size does not enter into the site comparison process, yet the variation in reactor sizes can be greater than the variation in population densities at roughly similar sites. (vi) The criteria imply a correlation between individual risks and site characteristics. (vii) The safety analysis assumptions are suspect – they refer to a ground level release of volatile fission products in semi-standard proportions with a release time measured in hours: such an accident is probably more relevant to a fuel handling incident rather than a real reactor accident. (viii) Even though the risks are small there are ethical grounds to support the principle that an individual who has not been offered any choice in the matter should not be subjected to a greater risk because he happens to live near to a reactor: this principle is negated somewhat by the use of population density based siting criteria. (ix) Modifications are needed to allow for different types of release, the inclusions of the effects of an emergency plan, the need to consider the corrections to allow for multiple reactors on the same site, the incorporation of local rather than national age-sex profiles or the use of age disaggregated population data, and a fundamental re-examination of the dose-casualty relationships since total casualty numbers involve considerations of very large populations exposed to low doses in addition to the possibility of very high doses to a few; and (x) only derived standards are specified, without an underlying numerical standard of what is an acceptable risk for members of the public, it is impossible to evaluate site ratings or even emergency plans in a meaningful way.

The solution to many of these problems is, according to Adams and Faux (1969), simply to remove the relaxed siting criteria and base siting on two critical areas: an exclusion zone which would be included within the station boundary and an emergency action area the size of which would reflect the acceptable annual expectation of casualty numbers. A combined assessment could be made of site and reactor safety using the same probabilistic assessment methods. This logic is implicit in Farmer (1967). The result would indeed be to remove all siting restrictions because the predicted death or adverse health effects probabilities would be so incredibly small as to make all siting measures unnecessary. The counter-arguments are presented in a later chapter.

Many of the criticisms about the relaxed siting criteria result from attempts to attribute the criteria an explanation in terms of the risks of nuclear power. An alternative view might be that they provide a way of comparing the population density and distribution characteristics of various sites with a normative standard based on the assumed acceptability of Hartlepool and Heysham. As such it seems quite reasonable to weight the sector and distance band distributions of population by weights which are some negative exponential function of distance. The question as to whether these weights are meaningful representations of the geography of the risk of nuclear power is an interesting one. Many opponents to nuclear power would argue for a flat function. Even those who favour nuclear power might wish to continue the weighting functions beyond 20 miles and argue about what form the numbers should take – for example, annualised individual death risks or maximum casualties for a worst case reactor accident. These and other suggestions are considered later.

Application of the relaxed siting criteria to the UK

As might be expected, the results of using the relaxed siting criteria depend on how they applied. If the most flexible, or relaxed, interpretation is used then none of the existing UK nuclear power station sites violate the criteria. That is to say, the total site and sector risk factors when accumulated over the 7 distance bands do not exceed the 20 mile limiting values given in Table 4.4. Indeed none of the sites violate the criteria if the summation is restricted to the first five distance bands, the 0 to 5 mile distance. On the other hand, if a more strict interpretation is adopted then two of the existing sites seem to violate the criteria. That is, the site or sector risk factors for certain distance bands exceed the limiting values in Table 4.4.

The Hartlepool site violates a strict interpretation of the relaxed siting criteria because there is too great a population concentration in the 3 to 5 mile distance band. The band limit that may be inferred from Table 4.4 is 111,789 people whereas the estimated population in 1971 was 209,042 and in 1981 either 138,457 if census enumeration district data are used or 145,378 if approximate 1km grid-squares are employed. Indeed the CEGB themselves estimate that there are 168,500 people living in this distance band; see Chapter 5 for further details. It would also seem that in 1971 some of the 2 to 3 mile sectors violated the sector limits. For example, the maximum population 30 degree 2 to 3 mile sector had a population estimated at 18,453 compared with the implied limit of 16,718. These sector violations had disappeared by 1981 mainly because the settlement concerned was demolished because of risks from nearby chemical storage tanks. However, all these violations disappear if a more relaxed interpretation is used because the excess 2 to 3 and 3 to 5 mile populations are more than compensated for by less than maximum populations in all the other distance bands. Nevertheless, it is hardly a satisfactory state of affairs that so much excess population should be located in what must be a prime evacuation area should a major accident ever occur. It may be that the full significance of this excess population, or even its existence, was not recognised at the time Hartlepool was selected.

The only other UK semi-urban site at Heysham also seems to be a marginal location with respect to the relaxed siting criteria. It satisfies the criteria for 1971 and 1981 using area pro-rated estimates based on 1km grid-square population data; see Chapter 7 for a discussion of different population estimation methods. However, it fails the criteria if 1981 census enumeration district data are used. Indeed the site here is so marginal, or conversely the relaxed criteria so finely tuned, that the extent of the failure increases dramatically if the coordinates used to define Heysham are shifted from either a point midway between the A and B stations to a location based on the A station reactors, or if a different definition of census population is used. In all cases the site only fails the criteria if an examination is made of all possible 30 degree sectors in one degree increments. It may be due to data estimation errors but it is certainly a most marginal site. It can be made to satisfy the criteria either by the selection of a 30 degree sector with the maximum 20 mile population or by resorting to the relaxed interpretation of the criteria.

Given the very relaxed nature of the current demographic criteria for nuclear sites it might well be thought foolhardy to push the criteria to such limits and select what seem to be the extremely marginal sites of Hartlepool and Heysham. It may be that the situation only arose because of gross estimation errors in the mid 1960s when accurate population counts may only have been available for the 0 to 3 mile distance band and when sector rotation was probably out of the question when manual methods were used. However, it is worth remembering that these sites were in fact declared acceptable by Ministerial fiat and predate the relaxed siting criteria which were supposedly tuned to ensure their acceptability. It would appear therefore that the most likely explanation is related to the problems of obtaining accurate population data for small areas at a time when census data was only available for large areas.

Whilst it is slightly astonishing that such marginal sites should have been selected, it is perhaps more surprising that Heysham should have been chosen as the site for two more AGR reactors in the late 1970s. Hartlepool it seems may not be extended and it now appears to be viewed, unofficially at least, as a major mistake not on demographic grounds but because of the close proximity of chemical industries and oil storage tanks. Additionally, it is even more astonishing that the CEGB were able to claim in the early 1980s that they face a shortage of suitable sites;

it is astonishing because really the relaxed siting criteria are hardly restrictive. It would seem that the CEGB still have no clear idea of the distribution of potential nuclear power station sites that satisfy the current demographic constraints.

TABLE 4.7 *Areas of the UK excluded by various relaxed siting policies*

Policy variant	Coastal sites	Sites %	Total land excluded	
			Area	%
I	2324	17.0	39159	17.1
II	340	2.7	13025	5.7
III	1732	14.0	29865	13.0
IV	1898	13.9	49130	21.5
V	3644	26.7	70053	30.7

Table 4.7 reports the results of applying various versions, or interpretations, of the relaxed siting criteria to both the coastal areas of the UK and to the total land area. Five different interpretations are evaluated. Policy I is based on the original Ministerial announcement of 1968, which may be taken to imply a limit of 500 people within any 30 degree sector within 1 mile of a reactor site. Policy II seems to be what is used and is based on a check of cumulative site and sector risk factors at 20 miles; this is the so-called relaxed or flexible interpretation. Policy III is the more strict version of policy II, there is check on the site and sector risk factors for each of the 7 distance bands. Policy IV is a version of policy II which uses one-third of the limiting values and corresponds to what the CEGB seem to use as a definition of a 'remote' site. Finally, policy V is similar but checks the cumulative site and sector risk values for each distance band. The effects of the different interpretations are quite marked; compare policies II with III, and IV with V. Nevertheless, the general impression gained from Table 4.7 is that there is no shortage of potential nuclear power station sites in the UK.

Figures 4.3 to 4.7 give a graphic impression of the effects of the different siting policies and show the locations of those 1km grid-squares where demographic factors would probably preclude nuclear power stations. Apart from the major urban areas, there are no real restrictions of any particular note. The results can also be reported by counties; see Appendix 2. These analyses would indicate that there are absolutely no grounds for believing that there is any shortage of sites in the UK using the current

100 Km

FIGURE 4.3 Infeasible sites in Britain according to a strict interpretation of
the 1968 Ministerial announcement (policy I).

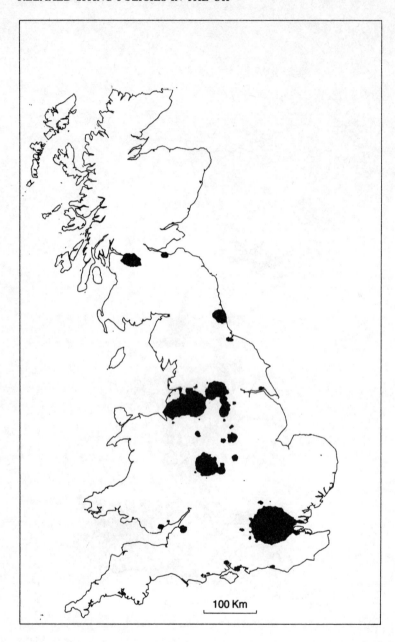

FIGURE 4.4 Infeasible sites in Britain according to the NII's relaxed siting
policy (policy II).

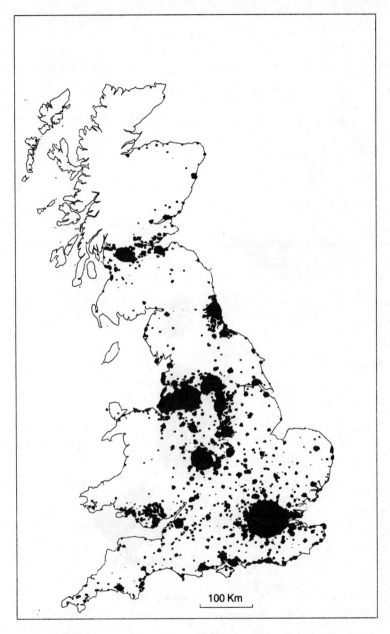

FIGURE 4.5 *Infeasible sites in Britain according to a strict application of the NII's relaxed siting policy (policy III).*

FIGURE 4.6 Infeasible sites in Britain according to the CEGB's remote siting criteria (policy IV).

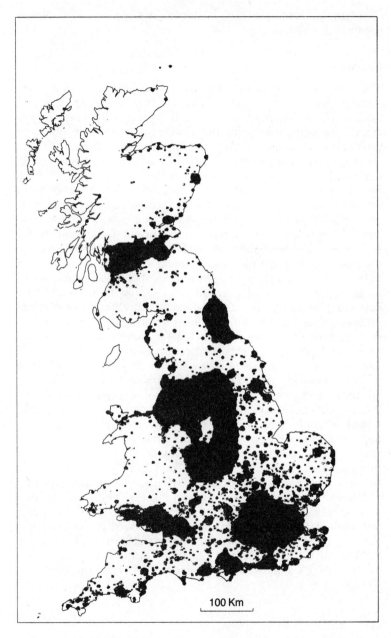

FIGURE 4.7 Infeasible sites in Britain according to a strict application of the CEGB's remote siting criteria (policy V).

demographic criteria; nor is there any particularly uneven regional distribution.

Conclusions

There can be little doubt that the UK's current relaxed siting demographic criteria have a minimal impact on the siting of nuclear plant. Whether of course the potentially suitable sites from a demographic point of view are also suitable on other grounds is difficult to determine. However, claims for site shortages would appear to be extraordinary. What it must reflect is the difficulty of finding sites in those particular areas preferred by the CEGB based on other grounds. This is probably a reflection of marginal cost economic factors, engineering factors, and possibly self-imposed environmental constraints. Nevertheless, there would now appear to be prima facie grounds for requiring a far more elaborate evaluation of alternative sites. In the past the shortage of sites argument has been used to justify both the need to develop a particular location and the absence of a detailed evaluation of alternative sites. Now there should be a demand that there is a detailed comparative evaluation of all potential sites that are not precluded on absolute grounds (either demographic or engineering) with a far greater input from both public and politicians. This is after all what happens in the US and there is no real reason why a similar comparative evaluation process cannot be used in the UK. Of course it will be resisted by all those concerned on the grounds of expense, complexity, delays, and necessity. Such complaints are due to the inertia resulting from three decades of complacency and pragmatism, and muddling through. The public, however, may well expect that *only* the best sites from the set of all available sites are brought forward to be developed. The CEGB would claim to do this but one may well wish to question their past performance, efficiency, comprehensiveness, rigorousness, and their reliance on manual rather than state-of-the-art computer-based site search and multiattribute evaluatory methods. Indeed it would appear that the basic approach has changed but little from that used in the late 1950s by the Central Electricity Authority (predecessors to the CEGB) and explained to Parliament (Hansard, 1957; pp. 678-83).

Appendix 2

TABLE 1 *Coastal areas of UK counties excluded by various relaxed siting criteria*

Region County	I	II	III	IV	V
Scotland					
Borders	25	0	5	0	30
Central	43	0	40	84	100
Dumfries & Galloway	3	0	1	0	6
Fife	48	1	29	58	75
Grampian	25	0	12	5	30
Highland	2	0	1	0	3
Lothian	41	12	33	27	47
Strathclyde	7	0	4	4	9
Tayside	38	0	29	20	54
Islands	0	0	0	0	1
East Anglia					
Cambridgeshire	100	100	100	100	100
Norfolk	36	0	26	10	49
Suffolk	27	0	15	8	42
East Midlands					
Derbyshire	100	100	100	100	100
Leicestershire	100	100	100	100	100
Lincolnshire	4	0	1	0	4
Northamptonshire	100	100	100	100	100
Nottinghamshire	100	100	100	100	100
North					
Cleveland	55	12	54	78	95
Cumbria	18	0	12	4	34
Durham	86	0	86	100	100
Northumberland	37	0	30	23	44
Tyne & Wear	96	91	100	100	100
North West					
Cheshire	49	58	78	100	100
Greater Manchester	100	100	100	100	100
Lancashire	43	2	44	70	83
Merseyside	71	56	76	100	100
South East					
Bedfordshire	100	100	100	100	100
Berkshire	100	100	100	100	100
Buckinghamshire	100	100	100	100	100
East Sussex	60	4	49	41	69
Essex	31	10	28	45	61
London	78	100	100	100	100
Hampshire	65	18	64	90	93
Hertfordshire	100	100	100	100	100

TABLE 1 *cont.*

Region County	I	II	III	IV	V
Isle of Wight	41	0	16	25	55
Kent	44	5	37	58	78
Oxfordshire	100	100	100	100	100
Surrey	100	100	100	100	100
West Sussex	66	0	44	68	92
South West					
Avon	52	5	45	85	97
Cornwall	20	0	11	5	35
Devon	36	0	26	14	55
Dorset	36	0	29	23	61
Gloucestershire	11	0	5	13	19
Somerset	22	0	14	0	34
Wiltshire	100	100	100	100	100
West Midlands					
Hereford and Worcester	100	100	100	100	100
Salop	100	100	100	100	100
Staffordshire	100	100	100	100	100
Warwickshire	100	100	100	100	100
West Midlands	100	100	100	100	100
Yorkshire & Humberside					
Humberside	25	1	23	24	54
North Yorkshire	23	0	20	6	38
South Yorkshire	100	100	100	100	100
West Yorkshire	100	100	100	100	100
Wales					
Clwyd	81	0	56	54	94
Dyfed	12	0	7	2	19
Gwent	7	0	15	100	100
Gwynedd	17	0	6	0	25
Mid Glamorgan	27	0	22	9	40
Powys	100	100	100	100	100
South Glamorgan	45	6	54	66	86

TABLE 2 *Total areas of UK counties excluded by various relaxed siting policies*

Region County	I	II	III	IV	V
Scotland					
Borders	3	0	1	0	5
Central	14	0	9	35	39
Dumfries & Galloway	2	0	0	0	4
Fife	27	0	17	42	56
Grampian	3	0	1	1	6
Highland	0	0	0	0	0
Lothian	20	3	17	31	36
Strathclyde	11	4	9	17	23
Tayside	5	0	3	3	10
Islands	0	0	0	0	1
East Anglia					
Cambridgeshire	19	0	9	4	30
Norfolk	12	0	4	2	19
Suffolk	14	0	6	2	24
East Midlands					
Derbyshire	32	4	21	74	76
Leicestershire	30	5	21	68	74
Lincolnshire	10	0	4	1	15
Northamptonshire	24	0	12	15	42
Nottinghamshire	41	12	34	77	80
North					
Cleveland	38	11	40	74	81
Cumbria	7	0	3	0	12
Durham	25	0	18	40	45
Northumberland	6	0	4	9	13
Tyne & Wear	86	72	90	100	100
North West					
Cheshire	30	11	28	74	77
Greater Manchester	88	98	100	100	100
Lancashire	38	25	40	70	74
Merseyside	84	84	94	100	100
South East					
Bedfordshire	36	3	20	57	70
Berkshire	38	14	32	56	65
Buckinghamshire	36	13	27	49	62
East Sussex	30	2	21	26	50
Essex	38	21	35	51	66
London	96	100	100	100	100
Hampshire	28	3	20	42	53
Hertfordshire	53	48	61	92	93
Isle of Wight	20	0	14	19	39

TABLE 2 *cont.*

Region County	I	II	III	IV	V
Kent	32	9	25	40	60
Oxfordshire	25	0	13	19	45
Surrey	53	39	53	90	93
West Sussex	31	0	17	44	59
South West					
Avon	43	10	34	86	88
Cornwall	14	0	5	0	22
Devon	12	0	7	4	21
Dorset	19	0	11	10	26
Gloucestershire	20	0	10	10	32
Somerset	14	0	6	1	22
Wiltshire	16	0	9	9	30
West Midlands					
Hereford & Worcester	13	0	7	15	26
Salop	9	0	5	10	20
Staffordshire	37	21	34	81	83
Warwickshire	30	19	28	55	62
West Midlands	69	74	79	100	100
Yorkshire & Humberside					
Humberside	17	1	11	12	35
North Yorkshire	9	0	4	15	23
South Yorkshire	55	26	51	96	97
West Yorkshire	66	57	67	100	100
Wales					
Clwyd	12	0	7	12	22
Dyfed	6	0	2	0	10
Gwent	31	1	27	58	65
Gwynedd	6	0	0	0	8
Mid Glamorgan	58	1	47	89	92
Powys	3	0	1	0	5
South Glamorgan	44	15	43	63	77

5 Three case studies of UK nuclear power station planning

So far attention has been focused on general aspects and theoretical approaches to power station planning in the UK. Some of the more interesting problems and difficulties can be clarified by looking at specific instances of the theory and practice of nuclear siting policies. For these reasons three sites are selected for a brief closer examination: Sizewell because of worldwide interest in the Sizewell B inquiry; Hartlepool, which has the dubious distinction of being perhaps the most urban of any nuclear site in the world; and Druridge Bay which is a recently adopted greenfield site which would almost certainly be unacceptable for nuclear power developments were it located in the US. The question is how has the nuclear power station planning process operated with respect to these sites?

Sizewell

Historical aspects

Sizewell was one of the early MAGNOX sites. An application for planning permission was first made in 1959 and Section 2 consent and deemed planning permission were received in February 1960 without a public inquiry being held. A nuclear site licence was granted in April 1961 and the Sizewell A station was commissioned in March 1966. There are two 325MW(t) MAGNOX reactors on the site with a declared net capacity of 420MW(e) in total. Each of the twin reactors in the A station is one-sixth the power of a modern PWR and by modern standards this is a very small nuclear power station.

In October 1969 an AGR station immediately north of the existing station was given deemed planning permission. Although the AGR was never built, certain preliminary site works were carried out before the project was stopped in 1971; there were major engineering problems with the AGR design at that time. In 1973 Sizewell was selected as a site for multiple consents and planning permissions for any of the reactor types then under consideration. This was a device designed to speed up the planning process by giving permission for nuclear developments in advance of any firm application by the CEGB! A total of 2500MW(e) of capacity was given approval, and it could have been provided by any combination of SGHWR, HTR, LWR, and AGR units, depending on government approval. In 1974 the government selected the SGHWR system, and in 1975 consent was given for a Sizewell B station of 2500MW. The HTR and LWR applications were assumed to be withdrawn at this point. In February 1978 the SGHWR programme was cancelled before any construction work began and Sizewell was selected as the site for the first PWR to be built in Britain, a decision later confirmed in 1980 when the first public announcement was made. Applications for Section 2 consent and for deemed planning permission were submitted in January 1981. The proposed Sizewell B PWR would have a total electricity output of 1110MW(e). In July, 1981 the Secretary of State for Energy announced a public inquiry with a view to establishing facts about the need, safety, waste disposal, and local impacts of the development. This might well be regarded as a public relations exercise, since the generic consents given in 1973 would have still been considered applicable to the PWR. However, there is a tradition in the UK of holding a public inquiry for the first reactor of a new type as a means of establishing its safety for use under British conditions. It may then be used on other sites without any need for a repeat of the safety arguments, although the extent to which safety can be regarded as a generic principle has yet to be tested in the UK, mainly because under the 1960 Act it is the site and not the reactor that is licensed. Anyway, in January 1983 the inquiry began and it lasted until 1985. It is probably the most expensive public inquiry ever held.

The Sizewell B Public Inquiry will undoubtedly be the subject of many books in its own right. Suffice it here to say that many anti-nuclear and environmental groups refused to participate on the grounds that the decision to build was a foregone conclusion. In addition, whilst the CEGB were expected to spend between £10 and 20 million of public money in preparing their case, no public funds were to be available for any of the objectors. Indeed it is expected that by the time the inquiry is over the CEGB will have spent about £200 million (one-sixth the total capital cost) on advance ordering of components and final design work. They also, incidentally, have to pay the cost of the inspector and the assessors.

Selecting the Sizewell A site

For various reasons the process used to select the Sizewell A site is well-documented. It is probably the first nuclear power station where there has been sufficient time for a detailed pre-selection site investigation process to be applied; indeed some £100,000 in 1961 prices or 0.17 per cent of the total cost of the station was spent on site investigations (Gammon and Pedgrift, 1962). The general area was selected because it had no fuel resources, and it was ideally suited to supply power to London and the south-east – distances of about 70 to 120 miles.

The first step was a detailed search in 1956 of some 40 miles of coast north of Harwich; see Figure 5.1. Much of the coastline is low-lying marsh and broads protected from the sea by shingle banks. By eliminating the areas shown to lack various technical requirements and those too close to centres of population,

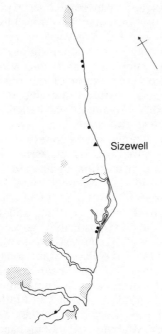

FIGURE 5.1 Search area for a nuclear power station site on the Suffolk coast and locations of alternative sites that were evaluated.

165

attention was focused on about a dozen localities on the Suffolk coast south of Lowestoft and north of Harwich. Precisely why this particular region of search should have been identified is not documented by Gammon and Pedgrift (1962) but presumably the close proximity of the 132kV national grid was an important factor.

A detailed desk analysis of the selected sites now followed, based on the Geological Survey, Admiralty Charts, Reports of the Commission on Coast Erosion, the County Development Plan and Survey, and aerial photographs. From this an initial selection of all likely sites was made and sites visited to find out more about recent developments and local features not otherwise apparent. Finally, seven sites in five areas were selected as sufficiently promising to warrant a more detailed investigation in 1957; see Figure 5.1. Gammon and Pedgrift (1962) write:'

> Each potential site was investigated in as much detail as was required in order to confirm its suitability and to develop outline civil engineering designs for the purposes of estimating comparative costs (p. 222).

These studies involved: hydrographic surveys and float tests to assess the ability of the sites to offer adequate mixing and dispersion of large quantities of warm water, and boreholes to investigate subsoil conditions. Other studies examined how the need for electricity supplies, water, and access could be met. Apparently 'The cost and extent of the offshore works of the cooling water system featured prominently in assessing the suitability of the sites' (Gammon and Pedgrift, 1962, p. 224).

Of the seven sites, two were being eroded by the sea and would require sea defence works, one was too high above sea level and would require considerable excavation for the turbine house, two required protection against sea floods as well as extensive piling, and two seemed suitable – Sizewell and Erwarton Ness. Sizewell had the advantage of close proximity to deep water, a relatively stable coastline and sea-bed, a low site but above flood level, reasonably good foundations, and reasonable road access. By contrast, the Erwarton Ness site was located on the banks of the River Stour some 5 miles from the sea, and it was probably too near to Harwich to justify contemporary restrictions on the population distribution as established by the remote siting policy of Marley and Fry (1955).

The Sizewell site was selected for further study although some doubts were expressed about the cooling capacity of the sea. It was feared that an offshore shingle bank might reduce mixing, and there was some concern about the foundation conditions of the low-lying land to the north of the proposed site which might be needed for possible future extensions. As Gammon and Pedgrift (1962) note,

> Following this additional work and after weighting the technical, economic, amenity, and other factors, the CEGB considered that the best site for a nuclear power station in Suffolk was at Sizewell (p. 230).

This is perhaps just as well, because it seems that the six alternative sites had been excluded on other grounds. Perhaps the deciding factor in favour of Sizewell was the need to construct an adequate cooling system which was both cheap and reliable.

Gammon and Pedgrift (1962) make no mention of the environmental aspects, except for some concern for the effects that different types of offshore works associated with the cooling water system might have on amenity. There is no evidence that any detailed environmental surveys, other than those for engineering purposes, were performed prior to the application for planning permission. It says much for the lack of interest in environmental matters that only a limited number of local objections were made to the proposed development of a nuclear power station in a designated Area of Outstanding Natural Beauty and none were received from any national amenity groups. It would appear that the East Suffolk County Council, the local planning authority who could have forced a public inquiry by objecting to the proposal, considered that Sizewell was the least objectionable of all the sites investigated from considerations of planning and amenity. They may well have done a deal with the CEGB. It was probably pointed out to them that the national interest required a nuclear power station somewhere on the Suffolk coast and that from a damage-limiting point of view Sizewell was the best site. If Sizewell was denied, then no doubt an alternative location in an even more environmentally sensitive area might then become essential. Yet the truth may be that in 1957-9 Sizewell was from a technical and economic point of view the only acceptable site to the CEGB. Today the situation has changed in that the CEGB now admit to an interest in Orford Ness, although again it seems that a deal has been made whereby no application would be made until the Sizewell site is fully developed.

Applications were made for consent in January 1959 and approval given in February 1960, without a public inquiry. It would seem that the invitations to tender were sent out prior to approval being given. The Sizewell A station was commissioned in 1966.

It seems that Sizewell established the basic principles of the power station planning process that is followed today with more detailed site investigations. The virtually total dominance of engineering factors is very noticeable.

Selecting Sizewell for Britain's first PWR station

Once a site receives a nuclear licence it seems that further nuclear developments are only a matter of time, unless there are good technical reasons against it. Sizewell is very well-placed to supply power to the south-east and is located in a region which is sufficiently sparsely populated to satisfy the government's remote siting criteria. It is also very hard to argue that once a site has gone nuclear further nuclear developments should be prohibited on planning grounds. In their evidence to the Sizewell B Public Inquiry the CEGB justifies the selection of Sizewell on the following grounds.

(1) Additional capacity is needed somewhere in the south-east and there are system benefits if suitable locations can be found in a power-deficient sector of the system (Arnold, 1982). That is to say, in order to support the system in the south-east power should be generated within that region, although sites outside but close to it could also be used. In the CEGB Statement of Case it is stated that:

> Studies of the operation of the system in the early 1990s show that some 70-80% of the optimum economic power transfers to the south are likely to be required in the south-east where there is a heavy demand for electricity. The studies also indicate that, under certain conditions, there is a possibility of the transmission capability to the south being inadequate and acting as a constraint on optimum economic operation (CEGB, 1982, p. 186).

This deficiency of power in the south-east was in part the result of the second oil price rise of the 1970s and the imbalance between coal based generation in the north and oil based capacity in the south, and the subsequent desire to minimise oil burn.

(2) There are three partially developed sites in the south-east with the potential for further nuclear development which were shortlisted – Bradwell, Dungeness, and Sizewell (Gammon, 1982, p. 6). These sites also have the advantage of being regarded as 'remote' and therefore satisfy the NII's requirement that only 'remote' sites be used for the first few PWRs, a type of reactor new to Britain. The possibility of selecting a greenfield site never seems to have been considered because of the CEGB's general

practice of developing existing sites where practicable; it might also have been regarded as having a far longer lead time with the prospect for considerable additional delays.

(3) Sizewell has two main advantages over Bradwell and Dungeness. First, because of the amount of preparatory and preliminary site works already carried out in the early 1970s. This was valued at £48 million and allowed a 6-month time advantage. Second, there is no need for additional transmission lines which would be required for the other two sites (Gammon, 1982, p. 6).

(4) Another reason for the CEGB's choice of Sizewell emerged during the public inquiry: this concerned their desire to reduce objections by selecting a site where most local people already accepted nuclear power. This is a new criteria for the selection of a nuclear site and it seems to have been a major factor in favour of the Sizewell decision. In addition, it was not anticipated that the local authorities in the area would object to the development proposals.

So it would seem that engineering convenience and other pragmatic considerations determined the choice of Sizewell for Britain's first PWR. Indeed, as was noted in Chapter 2, the choice of Sizewell was made without any comparison of the capital costs of developing alternative sites, there was no assessment of the net effective costs of the alternative sites, there was no detailed site comparison of the benefits and system operational factors, there was no comparative evaluation of landscape quality, visual impact, and ecology. It seems that these serious shortcomings in the CEGB's site selection and evaluation procedures were not regarded as being particularly relevant. The Secretary of the Board, Mr Baker, explained to the inquiry

I think you can see that these considerations are regarded by us as, if you like, endemic, underlying the way in which we do our business in site selection (Sizewell B, 1983, p. 62).

The response by Mr J. Popham of the Suffolk Preservation Society was to the effect that certain Section 37 duties had not been fulfilled until after Sizewell had been selected, but that the Section required these duties to be performed when formulating or considering sites. Popham concluded:

In fact the formulation process must include the original decision, because if you cannot compare the merits of sites which you are considering, then you cannot have regard to your duty (Sizewell B, 1983, p. 62).

The answer to this criticism was that the CEGB had a general consciousness of these items when formulating their proposals. In other words, the entire site evaluation and comparison process was largely based on informal argument, subjective opinion, and was partial.

It was subsequently discovered that the set of short-listed sites did not in fact include Dungeness, despite Gammon's evidence that it did. Dungeness was viewed as being unavailable because of potential site management problems. The short-list of sites was therefore: Sizewell, Bradwell, and Hinkley Point – the latter being in the south-west region of the CEGB, and mentioned only in internal CEGB documents. Its exclusion from Gammon's proof (1982) has been attributed to a concern that a potentially good alternative site not in the south-east region might weaken or confuse the transmission or system case for Sizewell. It would also probably require a detailed comparative assessment of costs and benefits of the form that did not exist when the original decision was made. In fact Popham accuses the CEGB of failing to fulfil their duty under Section 2 of the 1957 Act in that they admitted that in selecting Sizewell no economic evaluation had been made; no reference design was even available. Popham then tried to prove that Hinkley Point was a marginally better location on economic grounds but that it was excluded from consideration because internal Board documents indicated that it was being reserved for a commercial fast breeder reactor (Sizewell B, 1984, Day 247).

So it would seem that Sizewell was selected for Britain's first PWR because it was viewed as an easy option. The site was regarded as being environmentally acceptable (because the Countryside Commission did not oppose it) and the local people seemed to accept nuclear power. Safety matters do not seem to have been relevant; it was largely a matter of convenience and of engineering economics.

Alternative locations to Sizewell

The question arises as to whether or not pragmatism and convenience are themselves sufficient criteria for the choice of nuclear power station sites. When investments running into billions of pounds sterling are at stake it would seem prudent to consider the siting issue from first principles. If a site has to be found in Suffolk, or the CEGB's south-eastern region, and the arguments in favour of this degree of locational determinism are by no means that rigid, then it would be sensible to consider where are the feasible sites. This is important because the real issue is not simply the location of Britain's first PWR but whether Sizewell is a suitable location for two, possibly three or more, additional PWRs. The site was found suitable for a very small MAGNOX station and this should not automatically imply that it is also suitable for much larger stations without some comparative evaluation exercise to demonstrate the site's advantages in terms of environmental impacts, safety, and economics. So it would seem that the *real* issue has not yet emerged, because of the incremental nature of the nuclear planning decision-making process and its long historical timescale. In the same vein the concentration on a site's short-term advantages – such as no additional supergrid connections needed and compliant local opinion – should not be over-emphasised, because in the longer term the implications of development may be quite different. Sizewell C may well require an additional transmission line corridor which once created will itself justify a D and possibly further developments. However, the primary justification for the C station, should one be built, would be the presence of the B station! So it may be that the occurrence of chains of related decision-making extending over long time spans serves to hide the deliberate and systematically conceived planning that lies behind it all; or will increasingly do so in the future. If the question was whether or not Sizewell was a suitable location for the long-term development of several PWRs then the matter would be assessed very differently from that used when the question is is Sizewell a suitable site for a single PWR.

It is of academic interest to examine the Suffolk coastline to see whether potentially suitable sites can be found. From a purely demographic point of view large parts of the Suffolk coast, and indeed that of the south-east region of the CEGB, would qualify as feasible reactor sites. Figures 3.2, 3.3, 4.3 to 4.7 (pp. 120-1 and 153-7) show the location of infeasible sites and there are clearly large numbers in this area that should have been investigated in further detail; see also Appendixes 1 and 2 (pp. 122-5 and 159-62) which show results for coastal regions.

However, demographic criteria, whilst important, are not by themselves sufficient to identify potential sites. A study by Alderson *et al.* (1982) identifies a number of additional variables relating to a sample of 524 sites which were selected at 1km intervals along the entire coastline of the CEGB's south-east region. An attempt was made to identify whether each site was: already developed, protected either by ownership or by planning policies, classed as part of an area of special landscape quality (heritage coast, national or local nature reserve, site of special scientific interest), a tourist attraction, and suitable on physical grounds as a power station site. Table 5.1 summarises the results of these analyses and Figures 5.2 to 5.4 identify the sites that

171

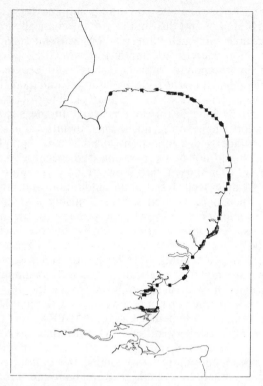

FIGURE 5.2 Sites identified on undeveloped land, that seem physically suitable, and satisfy the population criteria.

satisfy various combinations of siting criteria.

There would appear to be many potentially feasible sites in this region and it would be surprising if better locations than Sizewell in terms of safety and environmental variables could not be found. Indeed as Table 5.1 shows there seems to be a number of sites with potentially fewer casualties should a reactor accident happen, than Sizewell: see Openshaw, 1982b.

It is not seriously suggested that the methodology used here is sufficiently accurate and refined for use as a basis for actual siting decisions, but that it serves as an indication of the kind of assessments and evaluations that any reasonable lay person might well have expected the CEGB to have produced to justify the selection of Sizewell – or, indeed, for government to use when assessing the merits of nuclear developments at a particular location. It should be an integral part of the planning regulations for nuclear developments but at present it is not, and it may well be asked why the planning procedures for nuclear developments

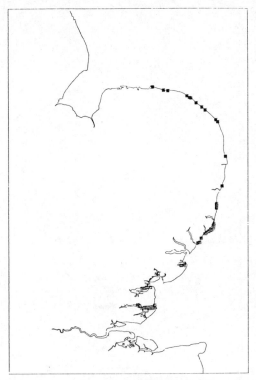

FIGURE 5.3 *Sites identified that also have no nearby beaches and no tourist facilities.*

TABLE 5.1 *Analysis of a sample of sites in the south east region of the CEGB in terms of restrictions on development*

Restrictions on development	Number of sites in sample of 524
Satisfy remote siting criteria	390
Satisfy remote siting criteria and not in any landscape designated area	45
Satisfy remote siting criteria and are owned by government	28
Already developed	140
Heritage coast	105
In areas of special scientific interest	141
Located within nature reserves	45

Source: Alderson et al. (1982).

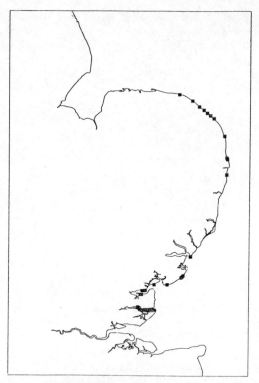

FIGURE 5.4 *Sites identified that also occur in areas of no special landscape designations or on National Trust land.*

are left to the developer to operate whereas in nearly all other areas of planning there are quite clearly defined goals that must be satisfied – for example, in structure planning with the compulsory public participation, evaluation of alternatives, and examination in public stages.

It is not difficult to imagine that the public might well be surprised if they were to discover that the 'best' sites were not being used. If the objective is to both protect the environment and minimise the hazards of nuclear power then there needs to be a very explicit and careful justification as to why these goals have been replaced by CEGB convenience and engineering factors.

But even an objective local determination of 'best' whether from a CEGB or the author's perspective is immaterial, if those sites fail to match yardsticks established at the national level. What desirable features should nuclear power station sites have?

If the justification for the selection of any site is the national interest argument then it is only fair that there exist a national interest definition of what is and what is not an acceptable site. These matters cannot and should not be left to the developer to determine under conditions of what amounts to almost total secrecy.

Hartlepool

This site is located on the edge of the major urban and industrial areas of Teesside, in north-east England.There is no documentary evidence to indicate how the site was selected and the manner by which it was approved is also largely based on hearsay, in the absence of any other evidence. Yet this site is probably the most urban of any nuclear power stations in the world, and it is one which is also in close proximity to the second largest petrochemical complex in Europe. It is seemingly a uniquely bad site from the point of view of public safety, and it is probably one of the few nuclear sites which will probably not be redeveloped – unless of course such developments occur after the oil industry has disappeared from the area about it.

It seems that the site was purchased by the British Electricity Authority (forerunners to the CEGB) in the early 1950s. The County Development Plan of 1954 allocated this land for use as a power station at some unspecified future date. The site was flat, it had nearby rail access and was well located to supply electricity to nearby industries. In 1961 the CEGB announced that the site would be developed in the period 1965-70. It was presumed that the station would be coal fired, supplied by the nearby Durham coalfield. However, in May 1966 the CEGB announced investigations into the possibility of developing the site for an AGR. A local newspaper announced that 'A £100 million nuclear power station may be coming to give a jet age boost to the Teesside boom'. In May 1967 a formal application was made for consent.

The application was particularly interesting as it was known that the site violated the current siting policy; see chapter 4 for details. Nevertheless, the development of an AGR station at Hartlepool was viewed as essential in order to gain export markets by demonstrating to the world that the AGR was suitable for urban siting. So although the site violated the existing siting criteria no one appears to have questioned this aspect. It seemed that the CEGB and the government were planning to develop a revised siting criteria which would declare Hartlepool to be an acceptable site, although the CEGB could not have known this when they submitted their application. At the time

175

the debate was concerned not with safety matters but with the economic merits of coal versus nuclear power stations. The National Coal Board viewed the matter as one of principle. If coal could not compete with nuclear power on the edge of a large coalfield then where could it compete? So Hartlepool was seen as a large psychological stake for the mining community. The CEGB, however, made it very clear that if Hartlepool was not to be nuclear then no station would be built there.

The debate about coal versus nuclear fuel costs is not particularly relevant, except that the Ministry of Power turned down an offer by the National Coal Board to supply coal at a substantially reduced price. What is more interesting is that the pro-nuclear lobby was sufficiently strong to persuade a Labour Government to make a decision in favour of nuclear power which was extremely unpopular in an area of traditional and substantial Labour support. It may only be a minor coincidence that the day after the CEGB made their application to build a nuclear power station the area concerned was transferred from the jurisdiction of Durham County Council (who would have opposed the nuclear development and forced a public inquiry) to the new and enlarged County Borough of Hartlepool which was known to favour the development of nuclear power as a means of increasing the civic status of the area and of reducing local unemployment.

In February 1968 the Hartlepool site was declared acceptable by Ministerial fiat, and was quoted as an example of the type of site now regarded as acceptable under a revised siting policy. The final decision had apparently been taken by the full cabinet with the support of the Prime Minister. The only subsequent contact between the CEGB and the local authority was over comments concerning elevations, landscaping, drainage, and access to the site which happened to be in an area of special scientific interest (Kendrick *et al.*, 1982).

An explanation of the revised siting criteria sent to Hartlepool County Borough stated that 'industrial development . . . is not seen as a determining factor . . . nor is it envisaged that the Minister would ask for restrictions to be imposed on industrial developments in its vicinity'. Indeed most of the subsequent oil and gas related developments were located outside of Hartlepool in neighbouring Teesside area. A planning committee minute of 17 April, 1968 notes that within specified distances the CEGB would require to be consulted over new developments '. . . as a means of maintaining public confidence'. This suggests a somewhat superficial view taken of the consultations by the CEGB at that time. The Nuclear Installations Inspectorate also used arbitrary safety zones of approximately 1 and 2 mile radii centred on the reactor. In the inner zone the NII should be consulted prior to the approval of any industrial and housing developments. In the outer zone they need only be consulted for housing developments involving the influx of more than 50 people.

The Hartlepool decision was certainly a controversial one and the real reason for its adoption is not known. It has also been suggested that the site was opposed by the Nuclear Installations Inspectorate, and the CEGB may well now regret the selection of this particular site because of its poor location. The principal siting factors appeared to be: the CEGB owned the site; the supergrid was nearby; there was a large local demand for power; and the site seemed to be suitable (despite the need for extensive infilling and massive concrete pillars 36 metres down into the underlying alluvial deposits to reach a rock formation). In addition, the local authority was strongly in favour and there were no public objections to it. The station was welcomed in an area where the traditional industries were declining. There was also strong nuclear industry support to demonstrate the superior economics of the AGR even on a coalfield site, and its suitability for urban siting. It would seem that the CEGB was made an offer they could not refuse: a turn-key deal for plant which would be cheaper than any other they owned! Sixteen years later the plant has not yet reached full output, and it is unlikely that it can ever make a profit because of the 10 years' delay in completion. In the US it would have probably been abandoned in the mid 1970s as hopelessly uneconomic because of the certainty of further massive cost over-runs. In the UK it is still regarded by the CEGB as a sound investment. If it works properly and the costing of the delays are ignored, then maybe it could one day be regarded in this manner.

Apart from being located close to major population concentrations, Hartlepool is also very near to a large concentration of hazardous industries. The siting of chemical industries in the vicinity was mainly a result of government-sponsored reclamation schemes in the early 1960s and occurred mainly after the nuclear station had been approved in the 1970s. The hazard aspects of these developments have always been overshadowed by the potential supply of new jobs from the new industries in an area of high unemployment. Certainly if most of the present chemical complex had been rejected on safety grounds because of the proximity of the nuclear plant, as perhaps it should have been, then the unemployment levels would be even higher. In any case, there were assurances prior to the nuclear plant decision that

177

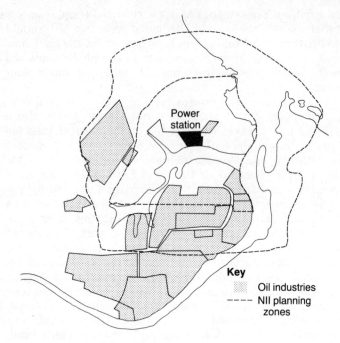

FIGURE 5.5 Location of Hartlepool Power Station and the proximity of chemical industries and related activities.

there would be consequent restrictions on industrial developments even though most of the petrochemical developments occurred in areas where additional residential developments would not have been allowed. Figure 5.5 gives a general impression of the proximity of the nuclear plant to the petrochemical industries.

The main concern is not whether the nuclear plant will affect the chemical industry but whether the latter could affect the safe operation of the power station. In this respect it is interesting to note that the Nuclear Installations Inspectorate has expressed strong concern about the effects of a gas cloud from the nearby chemical complex on the safety of the nuclear station. An internal CEGB document leaked to the press refers to the potential hazards from the adjoining petrochemical industry which have been introduced into the area following the initial granting of a nuclear site licence – although its development could hardly be regarded as either unexpected or surprising. It suggests

178

It is unlikely that a second station would be licensed without additional protection against gas cloud explosion and any such work could lead to pressure for retrofitting of equivalent protection to Hartlepool A (*Sunday Times*, 16 December, 1979).

This is often quoted as a prime example of negotiated safety or a lack of NII power to enforce important safety measures. The CEGB response has been to discount the problem pointing out that the leaked report is only a study of a theoretical risk of no practical consequence. Despite this, it is a valid comment to point out that there are large petrol storage tanks within about 1 mile of the reactor and an explosion there might be expected to cause damage up to about 3 miles away. Additionally, the International Atomic Energy Agency (1981) recommend that screening distances of between 5 and 6 miles should be used for sources of hazardous vapour clouds. If applied to the Hartlepool site this distance would encompass virtually all of the second largest concentration of petro-chemical industries within Europe (Chita, 1982).

The Hartlepool site clearly shows the difficulties of handling potentially large and hazardous industrial developments. The local planners do not have the expertise needed to offer any appreciation of the safety arguments; instead they rely on various government departments for advice. The planners see their task as ensuring that there is sufficient land for development and focus on the broader strategic aspects. Safety matters are the responsibility of government and the plant operators, both of whom adopt a policy of secrecy. At Hartlepool there was no published safety study, no justification for the selection of the site, and no assessment of possible consequences. Additionally, the scant local government files relating to the site were lost in 1974 when the administration of the area was reorganised. In some very real senses the main problem is that there is no legalistic formulation that is relevant to the siting process and without it, the reliance on informal and secretive methods can occasionally result in the most tremendous muddle.

Druridge Bay

This is a new greenfield site which is also located in North-east England. It may well be seen as a replacement for the lost development potential that may no longer exist at Hartlepool. This site is located in the middle of a 5-mile long sweep of beach and dunes which provide the first stretch of rural coastline north of the industrial areas of Tyneside. The CEGB announced its intention to investigate the Druridge Bay area for a nuclear power station in January 1979, following preliminary desk studies in 1976-7 and informal discussions with governmental agencies. It seemed a good prima facie case but it needed to be confirmed by drilling as well as by more detailed information from local authorities, statutory bodies, and voluntary agencies. It was also necessary to obtain a preliminary opinion on landscaping and to listen to any truly significant matters that may affect the site. A site was finally declared technically suitable for a nuclear station in 1980 and in August 1982 it was formally adopted as one of the possible sites for the third British PWR.

A public relations leaflet justifies the need for this site in the following way:

> To meet the demand for electricity in the 1990s and beyond the Central Electricity Generating Board will require new power stations in England and Wales. These stations will replace older and less efficient plant. . . . A new power station in the north-east of England is likely to be needed in this context. An alternative would be to build a new power station elsewhere and to construct a new long 400 kV overhead line to import electricity. However, this would not provide the same security of supply . . . (CEGB, 1980, p. 2).

Within the Druridge Bay area the Board examined various possible sites; these may be termed Druridge Bay north, south and centre and concluded that a central site would have the development potential for at least 2500MW(e). In fact there was only one alternative site of any merit and that was unavailable because of open-cast coal mining operations. Vague references have been made to other possible sites further north in areas of even higher amenity value and protected landscape qualities, although of course these would only be considered in detail if the preferred Druridge Bay site was to be denied to the CEGB – the same carrot and stick diplomacy that was applied at Sizewell. The CEGB also emphasised that whereas the site they bought at Druridge was of 700 ha – large enough for several reactors – it had at the time 'no plans to develop the site' (CEGB, 1980), by which it meant a firm commitment. Nevertheless, the CEGB decided that the next site to be developed after Sizewell B is to be Hinkley Point C followed by either Druridge A, Dungeness C, Sizewell C, or Winfrith; however, no firm decisions would be made until after the outcome of the Sizewell B Public Inquiry (Electricity Council, 1983).

Subsequently it came known that the CEGB were preparing

detailed engineering designs for developing the Druridge Bay site whilst at the same time saying in public that they had no such commitment, and indeed they had submitted no planning application. At this point, in 1983, the local planning authority (Northumberland County Council) decided it could no longer cooperate with the CEGB and ceased all negotiations with them. It was particularly annoyed that the Board had made no real case to justify departing from the established planning policies for the area which aimed at conserving its existing open character whilst encouraging recreational activities. Additionally, the CEGB had not examined the alternatives of a new coal fired plant at an existing site at nearby Blyth, or of redeveloping Blyth, or of importing electricity from neighbouring areas including Scotland. That the CEGB should be untroubled by such opposition is not surprising since it had a prima facie case for a nuclear development at Druridge and virtually the only grounds for its abandonment would be either a governmental order (most unlikely) or if the site was found to be technically unsuitable (also unlikely). Local political opposition is not something they have to consider.

Justification

The justification for the Druridge Bay site is basically that it is one of the few technically suitable sites (according to the CEGB) in the entire north-east region of the CEGB. It is not a protected area nor is it very close to urban developments. It is also conveniently located for connections to the supergrid. The CEGB would probably say that it is virtually the first technically suitable site you come across in the entire north-east region, if you start in Lincolnshire and head north. This seems to be based on the examination of aerial photographs of the coastline rather than on computerised site searches; indeed the CEGB appears to prefer the use of manual methods and to rely more on personal judgements than on computer analysis. The obvious question then is whether or not Druridge is really the *only* good site in the whole region. It may also be asked where the alternative sites were located, what sort of comparative evaluations were performed, and what hidden agenda of long term planning proposals exist for the Druridge site? It is unlikely that these questions will ever be answered in full.

The truth is probably that the CEGB identified the Druridge Bay site as looking promising early on during the 1960s, possibly as an alternative to Hartlepool. It offers a flat site, near to sea-level, in an area of low population and with planning controls on

181

all developments. The convenience of having supergrid connections close at hand may have itself been deliberate advance planning. So it may also be that the site was selected first and the justification made retrospectively.

The local needs and security of supply argument was always a weak one in an area which is rapidly de-industrialising on a massive scale. Electricity demand must have fallen dramatically with no prospect of any growth; at the same time the Hartlepool AGRs have not yet come into full operation. Indeed, it would now appear that the CEGB have dropped the local needs argument, preferring to rely instead on national power needs. The real justification for Druridge then is initially as a replacement for the coal-fired plant at Blyth which is due for replacement in the 1990s, and in an area of extensive long term coal mining operations such an admission would not be well received. The close proximity of the Cheviot Hills has led some commentators to suggest that Druridge will also become the site of a reprocessing plant with disposal depositories being constructed in the Cheviot granites; indeed the area has been considered as suitable at least for trial boreholes. Finally, it is observed that the convenience of having the supergrid in close proximity only applies to the first station on the site.

Environmental aspects

One advantage in selecting Druridge Bay as a site is that it is not designated as an area of outstanding natural beauty, nor is it a heritage coast, or even a site of special scientific interest. Instead it is an area where current planning policies permit no developments of any industrial or urban nature. It is an area of high amenity value. The CEGB argue that 'Care would be taken to ensure that a power station at Druridge blended well into the surrounding countryside' (CEGB, 1980). Sadly this is quite impossible. It is true the structure may be hidden from certain parts of the beach by high dunes (if they survive construction) but the area behind the dunes is a flat agricultural plain with long uninterrupted views from a considerable distance inland. From the point of view of minimising visual intrusion it is hard to imagine a worst location for a cluster of nuclear reactors. There is also the additional problem that even if the A station could be 'blended into the landscape' will the subsequent developments be so easily hidden? Once again, the initial application will be judged on the grounds that it is a single development whereas the site was probably selected mainly for its capacity for multiple developments which will not be considered relevant to the initial planning application.

Another advantage of the Druridge site is that there are nearby grid connections, and therefore the environmental damage caused by the need for additional power corridors does not occur – or does not occur initially at least. The question arises as to whether these good connections happen to be located nearby merely by chance, or was Druridge Bay always viewed as a possible site that would be enhanced if these connections were placed nearby? It would appear that this technical reason for the suitability of Druridge may have been created by incremental decision-making in anticipation of a future development there. It would tie in well with the CEGB's softly-softly, long-term creeping incremental approach to nuclear decision-making, in which the slowness of the decisions hides their ultimate impacts.

A CEGB perspective

Some aspects of the problems are clearly visible in a series of frank interviews with Mr Johnston, the CEGB planning engineer for the northern region, published in a local newspaper (*Sunday Sun*, 4 December, 1983; 11 December, 1983). He argues the case that the decision to build at Druridge, should it be made, will be based on a balanced judgement that best reflects national interest arguments. The CEGB regard their motives as being the motives of Parliament and the government of the UK. They are a public body set up by Parliament with directives given by Government, as such they have no will of their own although they are responsible for the siting decision. It's their responsibility for ensuring that people get an electric light when they put the switch on. They have a statutory duty to do this. Their opponents are therefore seeking to negate the will of Parliament and the Government (is it treasonable?). According to Johnston selecting a site is a matter of judgement and they try to reflect a judgement that is acceptable to the majority of the people in the country who of course do not live near the site in question. In short the CEGB aim for an informed layman's approach to attaining an acceptable balance based on opinion in general. Sometimes local government and local opinions have to be over-ruled in the national interest. He argues that it is not really a steamrollering process although it may look like that. The high success rate is merely a reflection of the CEGB's thorough and honest preparatory work and their will to succeed in the national good.

The combination of national interest and a degree of righteousness, makes the nuclear power station planning process virtually unstoppable. It is difficult to know what would persuade the CEGB to abandon the Druridge site they so clearly want. In 1984 they moved contractors onto the site in order to prove conclusively that the geology was suitable for a possible nuclear power station although it seemed to be related to the final aspects of design work for developing the site. There were fears that local anti-nuclear groups might have tried to stop this activity. The CEGB response was 'We do not anticipate that they would do anything to hinder the pursuance of our legal duty' (Mr Johnston, *Evening Chronicle*, Newcastle, 17 May, 1984).

The problems are clearly evident. Local people have no real power unless their views can be identified as being representative in some way of the majority of the people in the UK. Local public opinion polls, or national ones, are irrelevant because it can always be argued that any particular set of results are either unrepresentative, or biased, or a reflection of a misinformed viewpoint that would change given a better understanding of the situation. At any rate, the CEGB is responsible to Parliament and anything less than a direct government directive, supported by compensation for money spent already, would probably be doomed to fail.

The validity of the CEGB's approach would be greater if the 'balanced judgement' were to be based on an externally scrutinised scientific evaluation process. Natural justice suggests that the acceptability of the current process is greatly in doubt, and prior to any new sites being selected there really needs to be a major overhaul of CEGB attitudes, perspectives, and methods with respect to power station planning.

Do alternative, 'better' sites exist?

Very little detail is known about the CEGB evaluations that have resulted in the adoption of a site which could easily see the investment of tens of billions of pounds of public money. They have not named the alternatives that were investigated during the site search phase, in case disclosure unduly alarmed the public. The public and government are merely expected to accept that they got their locational decision right and, subject to approval by the Nuclear Installations Inspectorate, nothing more will be done. Druridge may well be a feasible site but is it a 'good' site, is it a 'best' site, is it a publicly acceptable site. What environmental trade-offs did the CEGB make? What detailed comparative evaluations did they perform? About these matters nothing is known. It would seem that secrecy is being used, once again, to hide the fact that the decision was made on the basis of hard engineering information and soft subjective opinions about everything else.

Some indication of the enormous amount of geographical freedom that exist in the Northern region is shown in Figures 4.3 to 4.7 (pp. 153-7). It would appear from Appendix 2 that one-half, and probably more, of the entire coastline in this region would satisfy the demographic criteria currently in force. There would seem to be many potentially suitable sites. Are the CEGB aware that so many demographically acceptable sites exist? The answer is probably 'no', although they would emphasise that other factors are involved, particularly environmental.

TABLE 5.2 *Landscape characteristics of a sample of northern region coastal sites*

Landscape designation	Percentage of sites
Area of outstanding natural beauty	13
Heritage coast	20
Site of special scientific interest	35
National park	8
Countryside park	0

A report by Lindsay and Norton (1981) gives some indication of the effects of various landscape or environmental constraints on the choice of reactor sites in the northern region. They studied a sample of coastal sites at 5 km intervals along the entire coastline. For each of this systematic sample of 107 sites they attempted to identify whether or not they were located within: areas of outstanding natural beauty, on a heritage coast, a site of special scientific interest, a national park, or a countryside park. The designations were only approximate because of the absence of accurate maps for the region. Table 5.2 shows the distribution of the sample of sites by type of environmental area. It should be pointed out that these landscape designations are by no means binding as factors that would exclude a nuclear plant. Additionally, many of the landscape designations overlap so that the best estimate of the areas likely to be considered unacceptable is considerably smaller than the 63 per cent figure often quoted by the CEGB.

The sample of sites is then tested against the current demographic criteria and the results crosstabulated against landscape designations; see Table 5.3. The results suggest that

185

TABLE 5.3 *Demographic and landscape characteristics of a sample of northern region coastal sites*

Satisfy demographic criteria	Landscape designations			Total sites
	AONB	HC	SSSI	
yes	22%	33%	40%	58%
no	0%	2%	25%	49%

AONB: area of outstanding natural beauty
HC: heritage coast
SSSI: site of special scientific interest

between 20 and 25 per cent of the northern region coastline might be considered suitable on demographic grounds, and lies outside the most sensitive environmental areas. Can there really be any sound justification for the CEGB only evaluating one site in this entire area? Are the hidden engineering constraints so binding as to exclude all others?

Given this amount of geographical freedom, it may well be worthwhile from a public relations point of view to deliberately look for the safest sites. Its not part of the CEGB's mandate to do this since the Nuclear Installations Inspectorate are nominally responsible for assessing site safety aspects. Nevertheless, an interesting question concerns how many of the sample of 107 sites would have lower casualties than either Hartlepool or Druridge, should a particularly severe reactor accident occur. Table 5.4 provides the answer; see Openshaw (1982b) for details of the accident scenarios considered.

It would appear that a number of the sites have lower potential casualties and are still satisfactory according to the demographic siting criteria. If the environmental constraints are excluded and the easiest sites physically are to be developed, then the best sites are in the north of the area; far to the north of Druridge. If the environmental constraints are included then the best sites are far to the south in Lincolnshire.

Conclusions

It would appear that the principal determining criteria are in practice factors of convenience to the CEGB. The largely hidden costs of supergrid connections, good as distinct from tolerable subsoil conditions, and undefined transmission system variables pertaining to some future forecast state would appear to be

TABLE 5.4 *Sample estimates of northern region sites with lower casualties than Hartlepool and Druridge*

Landscape constraints	Percentage of feasible[a] sites	
	PWR accident	AGR accident
Hartlepool site		
all sites	77	77
AONB excluded	56	56
AONB, SSSI, HC excluded	39	39
Druridge site		
all sites	93	89
AONB excluded	70	67
AONB, SSSI, HC excluded	44	44
Sites	58	58

Notes: a: satisfies demographic criteria
AONB: area of outstanding natural beauty
SSSI: site of special scientific interest
HC: heritage coast

dominant. Public safety is not considered a relevant siting matter except to the extent that a notional emergency plan must be considered operable and the site must satisfy the government's demographic siting criteria. There is no attempt to deliberately seek to minimise the health consequences of possible accidents through the use of distance as an additional safety factor, because everyone knows (or hopes) that major accidents cannot happen at CEGB-run nuclear stations. How do we know? They tell us that it is so.

This apart it would seem that the power station planning process has evolved over a long period of time but is in great need of a thorough overhaul. If the objective is to persuade the informed lay person that the best, balanced decisions are made, in the round as it were, then it really has to be put onto a far more scientific footing. There must be a detailed and explicit evaluation of alternative sites, both in terms of the engineering aspects as well as the environmental ones. These assessments should be published, for how else is the public to judge whether the CEGB have made the most appropriate decisions? There also has to be an explicit safety goal. It is not enough any more to simply delegate the site safety aspects to the Nuclear Installations Inspectorate. If sites are required in the national interest to satisfy a national demand for electricity, then the evaluations and site comparisons should also be done on a national basis rather than with reference to the supposed future peak hour flows between a meaningless set of arbitrary administrative regions which have little or no relationship to the patterns of demand, or anything else for that matter. Finally, it is a well-established principle of modern planning practice that there should be some scope for public participation in the decision-making. It is not at all satisfactory that siting decisions should be immune to local participation and views. Being informed is not enough if the information is subsequently ignored. This is important because in the long term the full utilisation of the potential of nuclear power will require that it has a high degree of public support and acceptance. The softly-softly approach is to keep quiet until nuclear power is essential, and without any alternative. This situation has not yet been reached and there is a real risk that in the short term a UK government will direct the CEGB to stop its nuclear operations because it feels that the majority of the public opinion may no longer support it. One way of reducing the prospect of this happening is to deliberately seek sites which are both technically feasible and publicly acceptable. At present it seems that technical optimality considerations dominate over all others for fairly marginal and undisclosed economic benefits.

In short there seems to be much that is bad about the CEGB's siting practices. By making 'balanced judgements' it is in fact neither protecting the environment nor safeguarding public safety to an acceptable degree. It is time that the basic principles underlying siting practices were debated in Parliament because of their great importance and implications for public safety should reactor accidents ever happen. If we wait until one or more reactor accidents have happened, very large sums of public money may well be at risk should sites have to be abandoned as unsafe. If the objective is to select sites based on balanced judgements then the balancing mechanisms need to be debated and made explicit.

6 Siting reactors in the US

Background

The siting and safety aspects of nuclear power stations in the US displays both similarities and major differences from that of the UK. One major difference is that the nuclear option is now regarded by a significant part of US society as the option of last resort. It seems that marginal economic savings are no longer so important provided there are other energy options. Wilbanks (1984) points out that there is an emerging consensus that environment and health should not be put seriously at risk by decisions, concerning nuclear power, and that it is not yet necessary to accept trade-offs of this sort to assure continuance of the American way of life. This situation could well continue into the early years of the next century. In the UK it is worth noting that a non-nuclear energy option is regarded as unavailable and there is every confidence that the PWR, which looks so risky in the US, can be made to work in the UK. The other major differences between the UK and US concern the historical development of nuclear power, the institutional organisation of power generation, the regulatory framework, and the extent to which the site selection process is discussed in public. Additionally, it seems that in the US there is an increasing move towards the siting of any new nuclear plant – and none is expected in the short-term – on isolated sites which have been explicitly chosen to mitigate the consequences of those accidents that cannot be handled by designed safety measures. This seems to be both a reflection of the need to reduce the residual risks presented by nuclear power, and of the expiry of the 1957 Price-Anderson Act in 1982 which hitherto had given limited liability to utilities in the event of reactor accidents.

Other reasons for the current lack of nuclear enthusiasm are

189

their poor economics especially in coalfield locations, less than expected load factors due to breakdowns and other outages, large cost over-runs, growing public opposition, uncertainties about the economics of reprocessing, uncertainties about the problems of high level waste management and storage, and the high cost of replacement power during shutdowns. It seems that these factors have a far greater effect in the US because in the UK it is the government and not private enterprise that is having to pay the bills. Additionally, Three Mile Island dealt the industry's public image a major blow from which it has still to recover fully. Other problems concern the decommissioning of obsolete plant, and the prospect of possibly more stringent discharge limits.

Despite the current difficulties, the US nuclear power programme started very well. The Atomic Energy Commission (AEC) became responsible for all aspects of atomic energy development in 1947. The AEC announced its willingness in 1950 to enter into study agreements with manufacturers, engineering construction industries, and electric utilities who wished to apply AEC reactor expertise to the design of civilian plants. Indeed by 1955, a total of 19 groups had made such agreements and 12 had evaluated reactor systems. At the same time the AEC embarked on a five year programme to develop five types of reactor that offered some promise for civilian power applications. One of these reactors was the 60MW(e) PWR built at Shippingport (1953-7). The PWR had been developed for use as submarine power units, and it was the only reactor in the early 1950s which was ready for large scale construction. By 1955 the AEC had several other reactor systems which had been developed to the stage at which it would be useful to build large prototypes. Accordingly a Cooperative Power Demonstration Programme, offering substantial governmental finance, was introduced to encourage the utilities to join the AEC and build prototype nuclear power stations.

The AEC offered assistance, waived use charges for the nuclear fuel, and provided training courses, while the power plant building firms offered loss leaders to convince the utilities that atomic energy had a future. However, none of these stations were expected to be competitive with coal fired plant and it was recognised that the major contribution to power supplies required fast breeder reactors (Paley, 1952). Nevertheless, this diverse approach contrasts strongly with the narrow focus of the UK reactor programme based on the gas-cooled graphite moderated system (Burns, 1967).

Despite the AEC support it seems that the utilities were not too happy about the prospects for nuclear power generation.

Major problems began to emerge when their liabilities in the event of an albeit unlikely accident were seen to be colossal. There was concern that the uncertainty about liabilities might deter utilities from building reactors. An AEC report suggested – on the basis of an almost total lack of actual operating experience – that the likelihood of a major accident was between 1 in 100,000 and 1 in 1000 million per reactor year, but that the consequences might well be in the worst case 3400 deaths, 43000 injuries, and property damage of 7000 billion dollars. This maximum credible accident assumed a 200MW(e) reactor located 50 km from a city with a population of 1 million. It envisaged a breach of the reactor containment with one-half of the inventory of fission products being blown by the wind in the direction of the city. The putative remoteness of the probability of such an accident occurring, juxtaposed with the astronomical numbers should it happen, created major insurance problems. The solution, as we have already seen, was the Price-Anderson Act of 1957 which offered utilities limited third-party liability in the event of an accident of 60 million dollars, and restricted Federal liability to 500 million dollars.

It seems that the combination of the Price-Anderson Act and a veiled threat that the AEC might itself enter the electricity business persuaded the major utilities to develop the first generation of reactors based on PWR and BWR designs; Dresden 1 (1955-59), Yankee Rowe (1956-60), and Indian Point 1 (1955-62) are one reflection of this change in attitude. However, it was not until the 500MW(e) Oyster Creek plant was ordered in 1963 (completed in 1969) without federal assistance that it seemed that nuclear power had made an economic breakthrough. This order marked the beginning of a flood of orders with a cumulative total of over 50,000MW(e) by the end of 1967. At this time the US had 28 times as much nuclear capacity on order as it did in operation. It is noticeable also that the average size of reactors increased rapidly from 315MW(e) in 1962 to 639MW(e) in 1965 to 1029MW(e) in 1969. This nuclear enthusiasm largely reflected the need for new sources of low cost energy, as demand increased exponentially at the same time as domestic fossil fuel prices rose steadily from the mid 1960s, and then more sharply after the 1973 Middle East policies on oil supplies and prices. Although most utilities seemed to be persuaded that the light water reactors were economically competitive by the early 1970s, costs were higher than predicted. Furthermore, reactor designs had not stabilised nor standardised, and there was unexpectedly strong opposition from environmental groups who claimed the light water reactors were dangerous.

The first order rush of 1965-8 had come before any plants above 200MW(e) had operated and was based largely on promise. The second and larger order rush of 1970-4 came as operating experience with large 600-800MW(e) plant was accumulating, but it was largely based on projections which took account of the early performance of the first few operational 500-800MW units. Nevertheless, there were tentative plans for the construction of a further 600 units in the next 25 years, resulting in 700GW(e) by the year 2000 or about half the total electricity needs of the US.

The fall in orders after 1975 was largely a reaction to a fall in electricity consumption – a traumatic experience for an industry used to an annual growth rate of 7 per cent (Burns, 1978). Demand for electricity increased by only 1.7 per cent in 1980, by 0.3 per cent in 1981, and −2.3 per cent in 1982. It reflected price rises, the increasing advocacy of energy conservation, and the increasing availability of new technologies with highly efficient energy use. The reduction in power demand meant that the utilities now had about 30 per cent more capacity than was needed, and partly as a result more plants were cancelled in 1982 than in any previous year. It seems that this was the final straw for many utilities which had been ordering plant in the 1970s on the basis of a rapidly increasing demand and with lead times of 10-15 years. The fall in demand greatly increased the financial difficulties of many utilities which were already in difficulties because of the virtual certainty of large cost over-runs on the nuclear plant still under construction but no longer needed, by regulatory changes in design criteria to cope with newly identified safety problems, by construction delays due to the faulty performance of contractors, by delays in obtaining price increases, and by delays in obtaining operating licences because of the successful stalling actions of interventors. One reflection of these problems can be seen in the increasing cost of nuclear plant; between 1963 and 1966 costs rose by 40 per cent; between 1967 and 1974 by between 300 and 400 per cent (Burns, 1978, p. 30). Another measure is the increasing time taken between ordering and commercial operation; in 1966 the average was 5 years, in 1970 it was 6.5 years, in 1971 9.3 years, and by 1976 it had increased to 12.3 years (NRS, 1977). Montgomery and Rose (1979) attribute these problems to the political, administrative, and judicial infrastructure that make up the regulatory process. Many of the difficulties resulted from the absence of standard designs, the trend towards progressive design changes, and the wide variability in the safety characteristics of different reactor types.

192

TABLE 6.1 *Annual reactor orders and cancellations in the US*

Year	Total reactors ordered	Net MW(e)	Operational reactor by order date	Nearly operational reactor by order date	Cancellations by order date	Date cancelled
1953	1	60				
54	2	450				
55	0	0	1			
56	1	175	1			
57	0	0	0			
58	1	65	0			
59	1	72	1			
60	0	0	0			
1961	0	0	0			
62	2	630	2			
63	5	3018	4			
64	0	0	0			
65	7	4475	7			
66	20	16426	19	1		
67	31	26462	25	3	2	
68	15	14018	9	5	2	
69	7	7203	3	2	2	
70	14	14264	7	6	1	
1971	21	20957	2	8	11	
72	38	41313	2	16	21	7
73	38	43319	0	12	28	0
74	34	40015	0	2	24	8
75	4	4148	0	0	4	9
76	3	3804	0	0	4	2
77	4	5040	0	0	4	9
78	2	2240	0	0	2	14
79	0	0	0	0		8
80	0	0	0	0		16
1981	0	0	0	0		6
82	0	0	0	0		18
83	0	0	0	0		3

Source: US Dept. of Energy (1982); Coffin (1984).

The combined effects of the changing fortunes of nuclear power can be seen in Table 6.1 which shows various time series relating to reactor orders, operations, and cancellations for the period 1953 to 1983. It is clear that most of the plant ordered in the 1960s has been built but over half that ordered in the 1970s has been cancelled mainly before construction started. Also only two new orders have been placed since 1978 and both were later cancelled. Currently no further orders are expected. As Robert Scherer, chairman of Georgia Power and head of the US Committee for Energy Awareness (a pro-nuclear group), said:

'No utility executive in the country would consider ordering one today – unless he wanted to be certified or committed' (Stoler, 1984, p. 9). It seems that a major cause of all the difficulties was the commitment to a large-scale nuclear programme before the nuclear industry had sufficient expertise, experience, and management skills to cope with the complexities of nuclear power. The enterprise was made even more problematic by a regulatory system based on everchanging regulations.

Currently about 12 per cent (55GW(e)) of total US electricity is nuclear generated. This quantity is expected to reach about 24 per cent by 1990. Unless there are some new orders soon, it seems that the 1990 level will also be characteristic of the early years of the next century.

Regulatory framework

From 1954 the AEC exercised two functions. It was charged with promoting nuclear power and it was solely responsible for establishing and enforcing safety regulations. To forestall obvious objections about bias and complicity, the AEC split into two sections in 1961: one concerned with operating and promotional functions, and the other with licensing and regulatory functions. This arrangement lasted until 1974, when the Nuclear Regulatory Commission (NRC) took over the latter functions.

Initially the AEC had little or no interest or responsibility for anything other than radiological effects and thus it gave no consideration to what may now be regarded as the traditional areas of regulatory concern, such as environmental impacts of accidents, or a concern for preserving the natural environment, as existed in the UK after the 1957 Electricity Act. Despite this narrow focus there was growing concern that the AEC took too long to give a decision on various proposals. In 1963 the average time taken between filing an application for a construction permit and the granting of a permit was 9 months; by 1970 it had increased to 20 months; and by 1976 it had reached 27 months and was still rising. According to Murphy *et al.* (1978) the main reason for the escalation was the impact of the National Environmental Policy Act (NEPA) in 1970, or as they put it '. . . in the expansive judicial construction of that vaguely worded statute' (p. 24). NEPA required the AEC to consider non-radiological effects on the environment previously deemed to be outside their jurisdiction, and would have had an effect on nuclear licensing. However, as interpreted by the US Court of Appeal for the District of Columbia in *Calvert Cliffs Coordinating Committee versus United States Atomic Energy*

Commission, 449 F. 2d 1109 (DC Circular 1971), its effect has been described as cataclysmic. That decision and its application to all pending nuclear projects brought AEC licensing activity to a virtual halt for over a year. It also required that the AEC publish for comment a set of realistic assessments of the environmental impacts of accidents at nuclear plants, and this probably gave the impression that nuclear power was somewhat risky.

These legislative changes seem to have coincided with a definite change in public attitudes to nuclear power. Up to 1971 most applications were uncontested and the public hearing both brief and harmonious. In short there was little opposition to either the development or the location of nuclear plant. But by 1971 every application was being contested and every contested hearing became a war (Murphy *et al.*, 1978).

A utility typically required dozens of permits before it could build and operate a nuclear plant. To some extent every situation was unique, and became a complex hodgepodge of federal, state, and local regulations. According to Minogue and Eiss (1976) the licensing process goes through an orderly chain of events commencing with the receipt of an application for a construction permit and continuing through the staff review of the Preliminary Safety Analysis Report (PSAR) and Environmental Report (ER), the Advisory Committee on Reactor Safeguards (ACRS) safety review, the public hearing, the awarding of a construction permit, the on-site inspection during construction and pre-operational and start-up testing, the staff review of the Final Safety Analysis Report (FSAR), a second ACRS evaluation, and finally the awarding of an operating licence for an initial period of 40 years (p. 85).

The applicant starts the process off by submitting three principal sets of information: a preliminary safety analysis report (PSAR), a preliminary environmental report (ER), and antitrust data so that the Attorney General can begin an antitrust review. As soon as an application is submitted all documents and correspondence become public information. According to Murphy *et al.* (1978) some idea of what happens is as follows although not necessarily in the correct chronological sequence.

Safety analysis reviews: Once a PSAR has been submitted the NRC staff evaluate the application in great depth including the site, location, meteorology, geology, hydrology, seismicity, reactor design, auxiliary plant, quality control, technical and financial qualifications of the applicant. The principal NRC regulations are contained in 10 CFR Part 50 and Part 100. There are a series of standard specifications of the information that have to be provided. When the evaluation of the PSAR is complete the NRC licensing staff issue a Safety Evaluation Report (SER). The application is now submitted to the Advisory Committee on Reactor Safeguards (ACRS) who decide whether the reactor can be operated without undue risk to the public. The outcome is presented in various correspondence and in a supplemental safety evaluation report.

Environmental report: The applicant submits a preliminary environmental report and this is reviewed in a similar manner to the PSAR, resulting in the production of an Environmental Impact Statement (EIS) which explores the need for the project and alternatives. At this stage other federal agencies are invited to express their opinions and give their approval to construction and operation. Although the NRC are dominant, several other federal agencies must also give their consent. The most important is the Environmental Protection Agency (EPA) who regulate non-radioactive chemical discharges and thermal effluents from power stations, it also regulates the design of cooling water intakes under the Federal Water Pollution Control Act of 1972. Pollution discharge limits may be set both by the EPA and the State; the latter are likely to be more stringent. A power plant application must also comply with state water quality standards.

The Department of Commerce may have plans for the management of coastal or lake zones and the plant has to comply with these. The Department of Transportation considers the effects on aircraft approaches. The Department of the Interior has a statutory duty for the protection of fish and wildlife values, while the Corps of Engineers require a permit to be obtained for any activity that may affect navigable waters and the disposal of dredged or fill material – and so on. On occasions over 21 federal agencies or departments have been required to issue permits.

In addition there may be various state legislation and regulation that affects the operation and construction of nuclear plant. Despite doubtful constitutionality, some states impose special restrictions – for example, California, Oregon, and Vermont. In addition federally delegated responsibilities for various aspects of the environment can influence siting. More-over, many states require a demonstration of a need for power and a utility will require a certificate of public convenience before a plant can be built. Finally, State Environmental Policy Acts (SEPAs) may require submission of environmental impact statements concentrating on various parochial aspects.

In addition, there may also be various municipal and county regulations that affect the licensing of nuclear plant. Some states have energy facility siting acts that provide for overriding local regulations.

What is even more interesting, is that since 1978 the NRC now routinely conducts detailed reviews of alternative sites and reviews the procedures used to select the proposed site; prior to that date they merely concentrated on the qualities of the proposed site (NRC, 1978). This is a most interesting difference from UK practices. Indeed the NRC rules require the explicit evaluation of alternative sites. The aim is to determine whether the candidate sites identified by the applicant are 'among the best which reasonably could have been found' within the region of interest (usually a utilities service area), although there is no definition of 'best'. Once it is determined that the alternative sites are among the best that could be found using reconnaissance level data, the NRC then compare the alternative sites to determine whether one of the alternatives is 'obviously superior'. A two-phase test is used: (1) considerations of water supply, water quality, aquatic biological resources, terrestrial resources, water and land use, socioeconomic and population factors to determine whether there is an environmentally preferred site; and (2) consideration of economics, technology, and institutional factors to determine whether, if an environmentally preferred site exists, such a site is 'obviously superior'. The applicant's proposed site would be rejected if it is found that an environmentally preferable alternative site exists that is obviously superior. Clearly the UK has something useful to learn here.

Public hearing: The Atomic Energy Act (1954) requires that quasi-judicial public hearings be held at one time before a single examiner but since 1962 before an ad hoc safety and licensing board (ASLB), consisting of two technically qualified members and one member qualified in the conduct of administrative proceedings. The objective was to ensure that the regulators had all the data they needed and that their recommendation was justified. The safety review and the environmental review are performed separately, and the hearing is conducted in two parts. People whose interest might be affected by the proposed plant have a right to intervene, ask questions, and submit evidence. By 1970 national groups of environmentalists were promoting and financing local intervention. Since the AEC defined the objects of the ASLB's proceedings in very broad terms, interventors could ask for documentary evidence of all the review process, subpoena participants as witnesses, cross-examine witnesses, present their own witnesses, and bring in outside experts. As a US government review put it, 'Almost anybody; for almost any reason, can hold up almost any reactor license for almost any length of time' (JCAE, 1971, p. 482). As a result the average time of hearings increased from 2 days in 1966 to 54 days in 1970, and later to 3 years.

If modifications were required to a proposal then there would be various negotiations between the applicant, the ACRS, and the AEC. Not surprisingly review and negotiation averaged 460 days in 1970 yet no construction work could be carried out until completion of the environmental stage of the hearings when site preparations might be started.

Operating licence review: After the public hearings a construction permit would normally be issued but before a reactor can be started-up an operating licence is required. The applicant submits a Final Safety Analysis Report (FSAR) which contains the final design and emergency plans. The NRC staff again prepare a Safety Evaluation Report and the ACRS make an independent evaluation. Although there is no legal requirement for a public hearing, a person may petition the NRC to hold a hearing when a similar pattern to the construction permit hearings is followed. An operating licence is normally granted for a period of 40 years, although NRC regulation and monitoring of the licensee continues throughout the life of the reactor.

Critical comments: Many aspects of this regulatory process have been criticised, but none more than the need for streamlining. Another major problem concerns the balancing of the impacts in different areas. When there is direct competition between competing public interests – for example, the need for energy and the preservation of an aquatic environment – the responsibility for this essentially political judgement has been given to the NRC and, or, the courts with the resulting uncertainty and delay. In particular Congress has been criticised for generally failing to state explicitly what is the acceptable balance of pollution abatement and social and economic disruption. As a result the EPA and the courts have been unwilling to force possibly serious economic dislocations (either by saying no to an application or forcing the use of a very different location) on the basis of the general need to balance health protection and economic impact. It might well be stated that in the UK this balancing is performed by the developer usually under conditions of total secrecy and with only the most general objective of a need to have due regard for the environment as a substitute for environmental impact statements. If the US system is slightly weighted against the developer, due to the complexities of treading the administrative path whilst facing the unintended cumulative impacts of diverse and independent statutes with diverse and redundant requirements and endless opportunities for litigation and delay, the UK system is heavily weighted in favour of the developer, although again there are opportunities for delay. This is perhaps not too surprising because the state is in fact both the developer, the examiner, and the statute enforcer in the UK. However, in both cases the task of achieving an acceptable balancing for the benefit of the public is exceptionally difficult. Unless and until a degree of quantitative measurement can be introduced, there will be no real solution.

What is also interesting is that the US nuclear industry would seem to prefer a UK style of 'federal' decision-making, whereas most opposition groups in the UK would like a US style with access to information and the prospect of being able to use the courts for litigation purposes. It may seem that the role of the interventor in licensing is unique to the US (Golay, 1980), where it accounts for much of the adversarial nature of the US regulatory process. Nevertheless, there are equivalents in the UK where interested persons may, within fairly well defined guide-lines, object at a public inquiry and present evidence and witnesses. However, most of the information which is public in the US about the details of the proposal and the safety reports are simply not available in the UK; one excuse is commercial secrecy, another is that there is no legal obligation for the information to be made public.

Most of the differences are a direct reflection of the fact that in the UK it is government which is the developer whereas in the US the utilities have been exposed to the full gamut of federal, state, and local legislation and decision-making. They also seem to have been unable to make use of national interest arguments to the same degree as in the UK and they face a far more strict application of radiological standards. It is very interesting that most of the first generation of US commercial reactors have now been shut down whereas all the UK equivalents are still running, having been given extended design lives. Nor does there appear to have been the same pressures for back-fitting existing plant as regulatory standards changed.

Other differences relate to the importance of evaluating alternative locations. In the UK this is not a problem because there are never any alternative locations to be evaluated. There is also no published or in-depth environmental impact analysis to confuse the issue. The CEGB have to satisfy various statutory duties and they claim to discharge these responsibilities in such a professional manner that there is no need whatsoever for additional planning studies of the sort common in the US. On the other hand, the detailed analysis of environmental impacts is seen by US utilities as a prime source of delays and uncertainties; the UK situation must seem like an impossible dream. In both countries there are perceived needs to speed up the planning process by resolving the safety issues in a generic fashion and

199

basically ignoring or reducing the emphasis given to such confounding factors as the environment. However, the permanency that is now seen to be associated with nuclear developments makes it important that the fullest and most detailed attention is given to all aspects of the initial siting decision, and to the public safety aspects of the first and all subsequent developments. There is a strong feeling that none of these objectives have been met in the past and are unlikely to be met fully in the future.

Early siting policies

The US has a very different set of siting policies than that found in the UK. The history is somewhat similar, although the response to the safety problems have been different and, one suspects more sensible, in the US. The Atomic Energy Act of 1954 made it unlawful to either construct or operate a reactor without a licence from the Atomic Energy Commission. The latter's jurisdiction was confined to matters of radiological health, safety, and defence matters. The Act allowed the AEC to develop whatever regulations and orders were thought necessary to protect health and minimise dangers to life and property. Accordingly the early military and research reactors were sited so as to use geographical isolation, in the form of a large buffer zone, to restrict the consequences of any accidents. Whether this policy was purely a reflection of the hypothetical dangers of reactor accidents or was based on real accidents is not known. Certainly the literature of the early 1950s in both the UK and the US gives the impression that a good deal of empirical information was available which indicated the need for reactors to be remotely located.

In the US a simple equation was used to specify the radius of the buffer zone (r) to provide the degree of geographical isolation considered necessary;

$$r = 0.01 \times SQRT \ (kW \ t)$$

where
kW t is the thermal power of the reactor, and r is the radial distance in miles (AEC, 1952).
This rule of thumb guideline had been developed by the Reactor Safeguards Committee, and related distance to nearest city to reactor size. It was related to the perceived consequences of a major accident and tried to ensure public safety by the provision of a buffer zone of such size that reactor sites would be isolated

from cities and other centres of population. It was probably derived empirically from studies of current reactor locations.

However as more and more facilities were being planned, so the policy of requiring the use of only the most remote sites was regarded as burdensome (Bunch, 1978). Indeed if these rules of thumb were to be applied to civilian reactors then they would preclude the nuclear option from large parts of the US, they would require sites far away from major load centres, and the utilities would have to purchase the very large areas of land to provide the necessary buffer zones. The power utilities naturally preferred sites near to major load centres and in the 10 years leading up to the interim siting guidelines of 1962 the importance of the buffer zone concept was greatly reduced for essentially pragmatic reasons. It is interesting that the equivalent buffer zone equation used by Marley and Fry (1955) in the UK as the basis for their remote siting policy predicted far smaller hazard ranges although the origins were probably similar.

The earliest commercial power application was made in 1953 and involved the construction of a PWR at Shippingport (Perry, 1977). The prevailing siting philosophy required an isolated area. However, the utility questioned the need for so much land; a buffer zone of 9-miles radius would have been required. The site it wanted was at Shippingport which is 20 miles from Pittsburgh and is located in an area of higher population density than the earlier reactors. In fact the site was selected for reasons other than site isolation; it was convenient for the utility and it was regarded as needed so as to maintain a US lead in the field of nuclear energy by exploiting the recent success of the PWR developed for the navy. Indeed as Bunch (1978) writes '. . . it was not apparent that any special consideration was given to site isolation in the selection process' (p. 3).

Indeed it was only during the construction that it was noticed that it was sited in a densely populated area, unlike all existing AEC reactors. The AEC countered these criticisms by postulating what is now known as the 'defence-in-depth' principle. A letter from W.F. Libby, chairman of the USAEC, to the Honourable B.B. Hickenlooper dated 10 March, 1956 sets out the basic logic. He wrote:

> . . . power reactors . . . will rely more upon the principle of
> containment than isolation as a means of protecting the public
> against the consequences of an improbable accident, but in
> each case there will be a reasonable distance between reactor
> and major centre of population.

This quotation is part of the now very familiar argument about

201

the risks of nuclear power. Regardless of the choice of location or the degree of isolation there is no such thing as an absolutely safe reactor. There is always present, regardless of the remoteness of its probability, a finite possibility of an event or series of events resulting in the release of unsafe amounts of radioactive material. The effects would depend on the population density of the affected area. If consideration were given only to safety factors then all reactors would be located in areas of the lowest possible population densities. However, in the early days it was widely believed that the growth and development of commercial nuclear power could not proceed under conditions of strict geographical isolation and that, in any case, its siting requirements were not significantly different from those of other industries. Nevertheless, it was recognised that a nuclear accident may be more severe than a conventional industrial accident and accordingly it was necessary to minimise the risks of such accidents happening and seek sites that would mitigate the consequences should they occur. This was to be achieved by: (1) recognising all possible accidents which could release unsafe amounts of radiation; (2) designing and operating the reactor to reduce the probability of such accidents to an acceptable minimum; and (3) by an appropriate combination of containment and isolation, protect the public from the worst consequences of an accident should one occur. Bunch (1978) provides a fuller historical interpretation.

This AEC view of balancing containment and site isolation represents a major turning point in siting strategies. Henceforth the policy of containment replaced that of isolation. This view was elaborated in 1960 when it was stated that

> . . . the relatively high close-in population density at the Shippingport site will require greater than usual reliance to be placed on the engineering safeguard features of the facility (USAEC, 1960, p. 9).

This philosophy of special compensatory engineering measures at densely populated sites later became a major feature of US siting practice, although it was present right from the beginning of commercial reactor siting. Indeed in 1955 an application was made for a 500MW(t) reactor at Indian Point, which is now one of the most controversial of all US sites. It is located 24 miles north of the New York City boundary. No special attention was given to the choice of site other than to note the degree of site isolation. There appeared to be 45,000 people within 5 miles and no consideration was given to the total population of New York (Bunch, 1978, p. 4). In 1965 and 1967 two additional units were located here. The question was not whether the site was sufficiently isolated but what design features were necessary to engineer the second and third units to make them acceptable at that site. There appears to have been little consideration of alternative sites or of the incremental risks posed by multiple developments. It was only later that population density constraints became important.

So it seems that the establishment, as a national goal, of a programme of developing the peaceful uses of atomic energy brought with it the need to site reactors close to the load centres they were meant to serve. This commercial need required that the AEC balanced its desire for site isolation, on grounds of safety, against the commercial needs of the power utilities. The result was an increase in the reliance on containment and other engineering safeguards and a decrease on the importance attached to geographical isolation.

The 1962 interim siting guidelines

In 1959 the AEC issued proposed general siting criteria (USAEC, 1959) which recommended that reactors should be located a minimum of 10 to 20 miles from large cities. In the same year a 60MW PWR demonstration reactor was proposed for a site at Jamestown in New York. The location of even a small reactor in a city location again demonstrated the basic conflict between utility preferences and the AEC desire for isolation. The application was rejected although the AEC still hoped that there might be a gradual evolution to metropolitan siting. The revised 1962 siting regulations were submitted as an interim measure, but have remained essentially unchanged since that time – although the NRC has issued additional siting related pronouncements in the form of siting decisions on specific cases, general design criteria, regulatory guides, standard review plans, licensing and appeals board decisions, and advice from the ACRS; all of which have contributed in a cumulative manner to NRC siting practices up to the early 1980s. Currently (in 1984) they are under review.

The AEC (1962) *Reactor Site Criteria* Title 10 Code of Federal Regulations Part 100 (10CFR100) sought a balance between site isolation and proximity to load centres, but it deliberately kept a degree of flexibility to allow for the evolution of the siting practices. It was regarded as an 'interim guide', because insufficient experience had been accumulated to permit the writing of more detailed standards which would provide a quantitative correlation of all the factors thought to be significant to the question of the acceptability of reactor sites. The details of distance requirements, whilst not mandatory, were regarded as

203

consistent with current siting practices. However distance requirements are only one aspect of the siting criteria. Section 100 10(b) states that the AEC will take: '. . . population density and use characteristics of the site environs, including the exclusion area, the low population zone and population center distance' into account when determining the acceptability of a site for a power reactor. Other relevant considerations include: the performance and experience accumulated elsewhere, the engineering safeguards, the inherent stability and safety features, and the quality of design, materials, construction, management and operation. While for any particular site, its size, topography, hydrology, meteorology, ease of warning and evacuation, and the thoroughness of the plans and arrangements for minimising injuries and interference with offsite activities are all important. A large number of design standards were also specified providing basic requirements that reactors had to satisfy.

The more interesting sections here concern the demographic siting criteria. This involved three components: an exclusion zone (EZ), a low population zone (LPZ), and a population centre distance. The geographical distances attributed to these three components reflected how well a reactor could handle a hypothetical maximum credible or design basis accident.

The exclusion zone: An applicant for a reactor licence is required to designate an exclusion zone with authority to determine all activities within it. The size is determined by the need to ensure that a person standing on its boundary for 2 hours immediately after a design basis accident does not receive more than a specified whole body and thyroid radiation dose. Transportation corridors are allowed to cross the exclusion zone provided they do not interfere with the operation of the reactor, and provided that appropriate control measures can be applied in an emergency. The objective was to control land use close to the plant, provide public protection in the event of an accident, and offer plant protection in the event of an offsite event. In practice the exclusion areas generally ranged from 0.1 to 1.33 miles with 0.4 miles being typical (Aldrich *et al.*, 1982). It may or may not be circular in shape.

The low population zone: An applicant is also required to designate a low population zone that immediately surrounds the exclusion area. The number of residents in the LPZ must be such that there is a reasonable probability that appropriate protective measures could be taken on their behalf in the event of a serious accident. In addition, the LPZ had to be of such a size that any person located at any point on its outer boundary who is exposed to the radioactive cloud resulting from the postulated design basis accident could not receive more than specified radiation doses. In practice this was to be achieved by the timely evacuation of the population from within this area. The LPZ might be thought of as a buffer zone with a typical radius of 2 to 3 miles located between the exclusion zone and large population concentrations in order to control or minimise the possible societal consequences of any accident.

The population centre distance: Finally, a proposed site will also have a population centre distance, defined as the distance from the reactor to the nearest 'boundary' of a densely populated centre of more than 25,000 residents, which is at least one-third greater than the distance from the reactor to the outer boundary of the LPZ. In practice, communities of 12,000 to 15,000 persons have been identified as the nearest population centre on the basis of projected growth. The definition of 'boundary' need not be merely political, it could also be demographic; and the specified distance might have to be greater should really large 'centres' of population be involved. The objective was to protect against excessive exposure doses of people living in large centres where effective protection measures might not be feasible. It was also meant to provide some additional protection for large numbers of people from accidents greater than those considered credible; the so-called class 9 accident.

Calculation of the distance factors: Figure 6.1 shows the relationship between the various distance factors. Their magnitudes can be computed from relationships contained in USAEC (1963). They can also be represented as the following approximate best-fit functions of reactor power

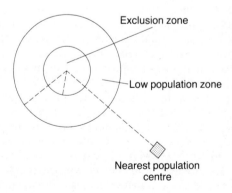

Exclusion zone

Low population zone

Nearest population centre

FIGURE 6.1 Areas and distances involved in the 1962 Reactor Site Criteria

205

$$r_1 = 0.2582 \times EXP\ (0.00088 \times MWT)$$
$$r_2 = EXP\ (-2.4671 + LN(MWT) \times 0.6937)$$
$$r_3 = 1.3 \times r_2$$

where
r_1 is the radius of the exclusion zone in miles,
r_2 is the radius of the low population zone in miles,
r_3 is the minimum population centre distance, and
MWT is the thermal power of the reactor in megawatts.
The distances given here and in USAEC (1963) are meant to be generally compatible with past siting decisions but they are only a guide. They can be adjusted upwards or downwards depending on the characteristics of the site or the reactor especially the containment leak rate and the nature of the engineering safeguards. As Bunch et al. (1979) emphasise it was apparent that the suitability of a site for any particular reactor cannot be determined from the guides in 10CFR100 either by calculation or by rule of thumb. The guides merely reflected past practice and current policy and their application depended very substantially on the detailed analysis of particular reactor designs.

Siting practice: The 1962 guides allowed plant design to compensate for the unfavourable physical characteristics of any site. The applicant was also free and even encouraged to demonstrate to the AEC the applicability and the significance of considerations other than those set out in the guides (USAEC, 1962). The extent by which the distance factor could be diminished is limited only by the degree of effectiveness attributable to the engineered safety measures and by considerations of proximity to very large cities. Indeed the effect of this flexibility was to allow reactors to be located closer to population centres than would have been permitted had there been strict adherence to the distance factors in USAEC (1963).

However, despite some flexibility all attempts to develop metropolitan sites failed on the grounds that they were seen to involve too great a reliance on engineered safeguards. In 1962 a 2030 MW reactor was proposed for Ravenswood on Long Island. It was twice the size of any previous reactor and there were 5 million people within 5 miles of it. The safety argument rested on a total reliance on engineered safety measures as a substitute for site isolation, while public fears focused on the fact that the possibility of an accident could not be completely eliminated. An application for a 600MW(e) reactor 9 miles from the centre of Boston was made in 1965; it too was rejected.

In 1966 a site at Burlington near Philadelphia, Trenton, and Burlington with a population within 15 miles of between 2 and 6 times that of Indian Point was rejected. Indian Point seems to have been taken as the limiting case until improved safety measures could be devised that would allow full metropolitan siting. In 1967 a site on Newbold Island within 5 miles of Trenton was also rejected; this is the period when equivalent UK sites at Hartlepool and Heysham were being proposed and later accepted. An application in 1967 for a site at Bailly within 10 miles of Gary was accepted as was an application for a site at Perryman which is 16 miles from Baltimore, although both were subsequently cancelled. So it would appear that in the UK reactors were being allowed on metropolitan sites with neither the benefits of geographical isolation nor secondary containment systems, and with complete reliance on engineered safeguards – something which was considered premature in the US.

The 1975 revisions

By the 1970s there was growing concern about the siting of reactors in densely populated areas, something which was not prohibited by the 1962 guidelines. This was reflected in the revised siting guidelines published in 1975 (NRC, 1975). It was also an attempt to develop generic guidelines based on a consolidation and standardisation of current practice as it was emerging from the results of various decisions. The situation was complicated by the large order rush and the wide range of different reactor types.

Regulatory Guide 4.7 (NRC, 1975) notes that based on past experience, the Nuclear Regulatory Commission (NRC) have found that a maximum exclusion boundary of 0.4 miles usually provides assurances that engineered safety measures can bring the calculated doses to within the 10CFR100 guidelines, even with unfavourable atmospheric patterns. Also based on past experience a low population zone (LPZ) of 3 miles is usually found to be adequate. These seem to be median values. Bunch *et al.* (1979) note that about 40 per cent of the current sites had distances of less than these recommended values and would have required compensatory design modifications.

There was also concern about the need to balance significant environmental, economic, and other aspects of alternative sites, including population. The 1962 guidelines did not properly include population density. There was a growing view that new sites should not be as densely populated as Indian Point unless there were other over-riding justifications. Although there is no specific regulation concerning population density in the vicinity of a power reactor site – other than that implicit in the EZ, LPZ, and population centre distances – the criteria published in Regulatory Guide 4.7 state that

> Areas of low population density are preferred for nuclear power station sites. . . . If the population density at the proposed site is not acceptably low, then the applicant will be required to give special attention to alternative sites with a lower population density (p. 16).

It was indicated that if the population density averaged over any radial distance out to 30 miles (i.e. cumulative population at a distance divided by the area at that distance) exceeds 500 persons per square mile at the projected time of initial operation or will exceed 1,000 persons per square mile during the lifetime (40 years) of the plant, then special attention should be given to the consideration of alternative sites with lower population densities. The idea was that these population densities would trigger an additional review of alternative sites, if none were found a site may still be acceptable (NRC, 1979). Bunch *et al.* (1979) observe that less than 10 per cent of the existing sites exceed the 500 persons per square mile density limitation and 2 exceed the lifetime limit; all were approved prior to the 1975 guidelines.

The recommended method for comparing the population distributions around sites is by computing site population factors (SPF). The SPF for any site is defined as

$$SPF_i(n) = (\sum_j^n W_j\, P_j)/(\sum_j^n W_j\, P^*_j)$$

where
n is the outer radius of the largest annulus located n miles from site i, W_j is the weighting factor for the jth 1 mile distance annulus (i.e. a radial distance band with an outer radius of j miles from the site), the weight is defined as a function of distance

$$W_j = r_j^{-1.5}$$

where r_j is the distance of the jth 1 mile distance band from the site (i.e. $r_j = j$), the W_j term is supposed to represent the approximate distance dependency of the short-term atmospheric diffusion factor;
P_j is the population of the jth annulus; and
P^*_j is the population of the jth annulus for a hypothetical population distribution with a density of 1000 persons per square mile, it is defined as

$$P^*_j = 1000 \times 3.14159 \times (r_j^2 - r_{j-1}^2).$$

The basic reference for the SPF is Kohler et al. (1974).

The SPF for a given bounding radius is different from the ratio of cumulative population densities within the same bounding radius, because of the distance weighting term. A densely populated area close to a reactor will have an effect on the SPF for all greater values of the bounding radii. However, with a simple density measure its effects will be progressively diluted. The SPF is intuitively sensible in that the influence of 100,000 people located close to a reactor yields a higher SPF than if it were located further away. The SPF also has a particularly simple interpretaton. For instance, an SPF of 0.5 for a bounding radius of 30 miles is numerically equivalent to that for an area having a uniformly distributed population density of 500 persons per square mile out to a distance of 30 miles. It is designed to allow a population distribution that is skewed in the radial direction to be compared with a uniformly distributed one.

The critical SPF (30) values of 0.5 and 1.0 may be taken as indicative of the population density criteria intended in Regulatory Guide 4.7.

The 1979 report of the Siting Policy Task Force

Despite the stop-gap measures provided by the 1975 revisions it was clear that the fluidity of the licensing policy made it difficult for utilities to forecast the acceptability of sites for nuclear plant. There was no absolute set of siting criteria but a whole range of site related pronouncements. To remedy these problems the NRC set up the Siting Policy Task Force (SPTF) with the objective of providing a single statement on siting policy and practice. Their recommendations were published in 1979 (NRC, 1979).

The SPTF make a number of recommendations concerning the development of siting criteria to be applied to new nuclear plants. It is particularly interesting that an attempt is made to re-establish demographic criteria as a reactor independent safety measure at a time when there is no real prospect of any new sites being developed. The SPTF establish *three goals* for siting policy: (1) Attempts should be made to strengthen siting as a factor in defence-in-depth by establishing requirements for site approval that are independent of plant design considerations. This is a reflection of the fact that current and previous siting policies have permitted plant design features to compensate for unfavourable site characteristics by improvements in design. This has had the

209

effect of reducing the role of site isolation as a safety measure.

(2) To take into account the risk associated with accidents beyond the design basis (class 9 accidents) by establishing population density and distribution criteria. Design improvements have greatly reduced the probability and consequence of design basis accidents to the extent that the residual class 9 accident risks now dominate safety studies. Whilst these risks cannot be reduced to zero, they can be significantly reduced by selective siting.

(3) To require that the selected sites will minimise the risk from energy generation. The selected sites should be among the best available in the region where the new generating capacity is needed. The siting requirement should be stringent enough to limit the residual risks of reactor operation but not so stringent as to eliminate the nuclear option from large parts of the country.

The establishment of these very worthy siting safety objectives is really a major advance, and is a good attempt to put siting criteria back on to a secure and defensible footing. The contrast with the current neglect of siting as a safety measure in the UK is really very striking. So long as the CEGB were building AGR reactors then the major differences in siting criteria could be put down to the inherently safer engineering of a gas cooled reactor. But when the CEGB are planning to build virtually the same reactor that is currently viewed as requiring a strengthening of the siting factor, then one may perhaps be forgiven for suggesting that UK reactor siting criteria reflect complacency and pragmatism rather than more fundamental safety objectives. The concern in the US with the need to mitigate the effects of the very unlikely class 9 accidents by siting measures is currently ignored in the UK.

The SPTF also identified a number of possible changes to siting policy to remedy the erosion of the original role of both population distribution and distance as twin elements of the defence-in-depth strategy which aims to provide an unquantified additional protection against the consequences of accidents beyond those for which the plant is designed. There seems to be a greater concern for class 9 accidents in the US. This seems to reflect a realisation that previous theoretical accident frequency predictions may either by wrong or subject to large margins of uncertainty or biased by the neglect of factors outside the designer's control; for example, human error and sabotage. The additional safety measures are to be provided by re-affirming 10CFR's use of population density and distance factors as elements of siting as was originally intended. The SPTF write

. . . this can be accomplished by isolating the plant design decisions regarding accident mitigation from siting decisions, by requiring fixed limits on population density and distances (NRC, 1979, p. 41).

Siting is also regarded as a means of reducing the residual risks by requiring the selection of sites with a minimum of unfavourable characteristics. The SPTF write

Reducing residual risks through conservatism in siting criteria should result in increased confidence that plant and site combinations result in reasonable assurance of no undue risk to public health and safety (p. 44).

The specific recommendations of most interest here concern these proposed binding siting criteria.

There are three principal recommendations.

(1) There should be a fixed exclusion distance based on limiting the individual risk from design basis accidents. A value of 0.5 miles was suggested as being a reasonable minimum.

(2) There should be a fixed minimum emergency planning distance of 10 miles. They write 'The physical characteristics of the emergency planning zone should provide reasonable assurance that evacuation of persons, including transients, would be feasible if needed to mitigate the consequences of accidents' (p. 46). Previously the LPZ was the area covered by the emergency plan and would usually be the same sort of area as covered by UK emergency planning measures.

(3) There should be specific population density and distribution limits outside the exclusion area. No definite values were given although some illustrative suggestions were made:

(i) from 0.5 to 5 miles the population density at the beginning of reactor operation should not exceed the larger of half the average population density of the region or 100 persons per square mile; the population should not be expected to increase to more than double the original population during the life of the plant, and no more than one-half of the allowed number of persons would be permitted in any single $22\frac{1}{2}$ degree sector;

(ii) from 5 to 10 miles the population density at the beginning of reactor operation should not exceed the larger of three-quarters of the average population density of the region or 150 persons per square mile and no more than one-half of the allowed number of persons in any single $22\frac{1}{2}$ degree sector; and

(iii) from 10 to 20 miles, the population desnity at the beginning of reactor operation should not exceed the larger of twice the average population density of the region or 400 persons per square mile, and no more than one-half of the allowed number of persons in any $22^1{}_2$ degree sector (NRC, 1979, pp. 49-50).

Beyond 20 miles the societal risks were considered too small for any specific population limits. The decision as to the precise numerical values and whether regional or nationwide population densities should be used was left for further evaluation.

There were 8 other recommendations which concerned such matters as: stand-off distances from hazardous locations, interdictive measures to limit ground water contamination, revised seismic standards, the monitoring of post-licensing changes in offsite activities, avoidance of sites that require special design changes to compensate for unfavourable site characteristics, a change in regulations to allow for site approval or disapproval at the earliest possible date, to allow state disapproval to terminate a proposal, and a common basis for comparing the risks for all external events.

The SPTF represented a compromise between safety and convenience. On the one hand they wanted the development of siting criteria that re-emphasised the contribution of favourable site characteristics in order to achieve a low level of residual risks; but on the other they emphasised that the siting criteria need not be so stringent as to eliminate large regions of the US from potential reactor siting. However, in general they took the view that the numerical demographic criteria should be more conservative than those at some existing already licensed sites. It was pointed out that these sites were not unsafe only that in the future best use should be made of sites with more favourable characteristics in order to provide additional levels of protection to public health and safety. The principal unresolved question concerned the value of the criteria to be used and it was recognised that this decision cannot be independent of other non-demographic siting factors.

Proposals for new siting criteria

In 1980 the NRC was directed to develop new regulations that established the demographic requirements for the siting of nuclear facilities based on the use of maximum population density and distribution criteria which were independent of the differences in design among plants. Furthermore, in developing these demographic criteria a full range of reactor accident scenarios, including those beyond the design basis were to be considered. Finally, the siting criteria must not preclude the siting of nuclear reactors in any region of the US; see Public Law 96-

295, 94 Stat. 780, June 30, 1980. Thus it seems that most of the principal siting recommendations of the SPTF have been accepted and further work authorised to set density and population distribution limits. The NRC also accepted the first two of the three siting goals used as a basis for the SPTF's recommendations (NRC, 1981). The third goal was modified to reflect the view that neither consideration of the risks of other power generating sources nor considerations of the preclusion of the nuclear option from any region of the country should supersede the basic responsibility to protect public health and safety. Three additional considerations were also added:

(1) The response of the NRC in reviewing an application with respect to siting criteria must be predictable and not subject to variability. That is the criteria must be both simple and specific.

(2) The criteria should not require extensive investigations and should be based on reconnaissance level information.

(3) There should be a reasonable assurance that public health and safety, both individual and societal risks, will not be endangered by the selection of any site under the criteria (NRC, 1981, pp. 6-7).

The NRC commissioned a series of studies to provide a technical basis for the subsequent siting criteria. It seems that the publication of the new criteria have been delayed pending a review of the severe accident source terms and the establishment of a safety goal. Meanwhile many of the analytical studies on which a decision will be based have been published. So although no final set of revised siting criteria are imminent (in June 1984) the various technical studies give a good indication of what the criteria will look like. These studies have been concerned with establishing the demographic characteristics of existing sites (Bunch et al., 1979) as well as examining the impacts of alternative demographic criteria (Aldrich et al., 1982). The latter cover those aspects considered relevant to siting policies: the consequences of possible accidents, the population distribution characteristics of existing sites, the availability of sites, and socioeconomic impacts. Additional demographic analyses were undertaken to supplement the site availability studies of Robinson and Hansen (1981), particularly interesting are Durfee and Coleman (1983) and Kelly et al. (1984) which look at the impact of a small number of alternative criteria. These demographic criteria apparently emerged from comments and discussions that followed the SPTF report and from the alternatives specified in NRC (1981) and Aldrich et al. (1982).

The concept of an exclusion area is retained and, according to Kelly et al. (1984), is fixed at 0.5 miles. The question as to

213

whether to use national or regionally variable population limits was resolved by setting national limits. It was considered that the critical test area is the north-east (defined as the area east of the 90th meridian and north of the 39th parallel). If demographic criteria can be selected in this region which adequately protect public health and safety and allow reasonable siting potential, then they should be capable of application in all other areas. It has also been decided to have both population density and limits on the populations allowed in various directional sectors; the latter is designed to limit siting near population centres. Finally, the question as to whether or not the criteria are to be absolute or a trip-level limit which would trigger a different mechanism for reviewing sites that fail the criteria, was left for a later decision. The opinion expressed in NRC (1981) was that the criteria should be absolute.

The NRC requested that a total of six different alternatives be tested. According to Kelly *et al.* (1984) there are three possible alternatives: limits that are more restrictive than present guidelines, limits which are a refinement of present guidelines, and limits which are significantly less restrictive than present guidelines. As no basis for any significant changes from present practice have been identified so it is concluded that the criteria should represent a refinement of the present guidelines. Table 6.2 gives the six alternative density and distribution criteria that were analysed by Kelly *et al.* (1984) for the NRC; alternative 4 seems to be the current favourite.

Each alternative has: (1) an exclusion zone with a radius of 0.5 miles; (2) a restricted population density area between 0.5 and 2 miles with a maximum population density of 250 persons per square mile; (3) a population control area of between 2 and 30 miles with specific population density restrictions, and (4) sector limits on any two adjacent $22\frac{1}{2}$ degree sectors which are designed to provide additional protection for large population

TABLE 6.2 *Six alternative US demographic criteria*

Area attributes	Alternative					
	1	*2*	*3*	*4*	*5*	*6*
Exclusion area radius (miles)	0.5	0.5	0.5	0.5	0.5	0.5
Density of restricted area	250	250	250	250	250	250
Density of population control area	500	500	500	750	750	750
Density of adjacent $22\frac{1}{2}$ sectors	1000	1500	2000	1500	2250	3000

Source: Kelly *et al.* (1984), p. 13.

214

TABLE 6.3 *Maximum allowable populations for the six alternative US siting criteria*

Distance band (miles)	Alternative					
	1	2	3	4	5	6
	Radial limits					
0-2	3,142			3,142		
0-5	36,128			52,622		
0-10	153,938			229,336		
0-15	350,287			523,860		
0-20	625,176			936,194		
0-25	978,605			1,466,289		
0-30	1,410,573			2,114,289		
	Cumulative population within two adjacent sectors					
0-2	785	1,178	1,571	785	1,178	1,571
0-5	9,032	13,548	18,064	13,155	19,733	26,311
0-10	38,484	57,727	76,969	57,334	86,001	114,668
0-15	87,572	131,358	175,144	130,965	196,448	261,930
0-20	156,294	234,441	312,588	234,048	351,073	568,097
0-25	244,651	366,977	489,303	366,584	549,876	733,168
0-30	352,643	528,965	705,286	528,572	792,858	1,057,144

Source: Durfee and Coleman (1983), p. 172.

concentrations. These criteria have been operationalised by Durfee and Coleman (1983) as involving a series of radial zones with radii of 2, 5, 10, 15, 20, 25, and 30 miles. The angular sectors are each of $22\frac{1}{2}$ degrees beginning with the first sector centred on the north direction. The population density limits in Table 6.2 can also be expressed as population limits; see Table 6.3.

Application of proposed criteria to the US

Kelly *et al.* (1984) provide details of the effects of applying the six alternative criteria to various regions of the US and discuss how the impacts can be measured in terms of site availability. Their results reflect work by Robinson and Hansen (1981) who used a 5km by 5km gridsquare data base for all of continental US to identify the effects of alternative siting criteria on site availability taking into account constraints such as the presence of restricted area, topographic restrictions, seismicity factors and water availability. A two-phase data analysis was performed. First those sites (i.e. 5 km cells) which were inherently unsuitable for nuclear

215

plant were excluded, and then the remainder tested against alternative demographic criteria. The results showed that many regions had minimal site problems in the western US but that some of the eastern states were sensitive to the choice of sector criteria. Kelly *et al.* (1984) extend this work and examine two alternative types of criteria: (i) uniform density criteria for the 2 to 30 mile region with a limit of 250 persons per square mile within 2 miles; and (ii) as (i) but with the additional limit of not more than one-eighth of the total population in any one sector. Table 6.4 shows the effects for the most affected western and eastern states. Whilst it may appear that there is a shortage of potential sites in some of the eastern states, there are in practice sufficient sites for nuclear power options up to at least 1999.

TABLE 6.4 *Effects of different population density limits on the percentage of land area available for siting nuclear plant in the US*

State	Uniform population density[a]			Uniform density plus sector[b] constraints		
	250	500	750	250	500	750
Western States						
Arizona	45	46	46	39	na	na
Idaho	53	54	54	45	na	na
California	53	57	59	37	na	na
Washington	55	58	59	37	na	na
Oregon	64	66	67	53	na	na
Minnesota	65	67	68	61	na	na
Utah	67	68	68	61	na	na
Montana	67	67	68	63	na	na
Eastern States						
Connecticut	4	28	45	.2	3	6
Rhode Island	4	24	43	0	0	0.8
New Jersey	13	26	32	4	5	10
Massachusetts	21	40	46	2	7	12
Maryland	47	58	66	14	21	24
New York	50	62	65	14	23	27
Ohio	51	69	75	9	21	25
Pennsylvania	53	67	72	14	25	30
Florida	58	65	67	26	40	45
Louisiana	62	65	66	28	38	42

Notes: [a] persons per square mile.
[b] not more than one-eighth of the total population allowed in any one sector.
Source: Kelly *et al.* (1984), pp. 23-4.

The six criteria of Table 6.3 have been applied to the 91 existing sites given in Aldrich *et al.* (1982) even though they would in practice only be used for new sites. Of these 91 sites, 11 exceed the Regulatory Guide 4.7 (1975b) nominal population density criteria of 500 persons per square mile out to 30 miles; all of which had been ordered in the 1960s. Seven of these 11 sites violated all the six alternative criteria, while only 3 of the remaining 80 sites violated all the criteria; 53 sites satisfied them all. This indicates that the population alternatives represent a refinement of current practice, as indeed was intended. It seems that the sector limits are the most binding, but even then the 11 sites that failed all the criteria would all pass alternative 4 if the sector limits were increased from 25 to 31 per cent of the total allowable population. It was concluded that even without this relaxation, 81 of the 91 existing sites would pass at least one of the six alternative criteria with 76 passing alternative 4. Moreover, 48 of the committed sites satisfy all the criteria and a further 15 only fail the most restrictive alternative. Furthermore, some ninety uncommitted sites meet all the demographic constraints and an additional thirteen only fail alternative 1. Kelly *et al.* (1984) conclude

. . . that application of any of the population distribution alternatives considered will not limit the availability of the nuclear option in the northeast quadrant of the United States. Furthermore, the analysis demonstrates that viable sites exist even in the states and service areas with the largest population densities (p. 71).

These conclusions are also supported by Durfee and Coleman (1983) who examined the impact of the six alternatives on three areas: the Pennsylvania, New Jersey, Maryland (PJM) area in the east; the Chicago centred region in the mid-west; and the Washington-Oregon area in the far west. They were selected to provide a meaningful cross-section of the US and a basis for understanding the overall restrictiveness of the population criteria in relation to urban centres. Table 6.5 shows the percentages of the state areas in each region that would be excluded as potential nuclear power station sites. Again it appears that there are many possible sites even in the urban north-east. The overall percentages of excluded areas range from 21 to 39 per cent for the north-east region, 12 to 32 per cent for the Chicago region, and 2 to 7 per cent in the far west region.

Durfee and Coleman (1983) also examined the effects on 248 individual sites which were broken down into 91 primary sites with active plants and 157 secondary sites where applications have

TABLE 6.5 *Percentage of land area in selected parts of the US excluded by various siting criteria*

State	% of area excluded by alternative					
	1	2	3	4	5	6
Connecticut	87	78	75	78	62	57
Delaware	32	23	20	25	19	17
District of Columbia	100	100	100	100	100	100
Maryland	52	45	43	46	41	38
Massachusetts	66	41	31	42	26	21
New Hampshire	17	15	8	15	9	8
New Jersey	84	75	73	74	68	66
New York	41	30	25	31	23	19
Ohio	55	36	29	37	26	21
Pennsylvania	40	29	25	30	23	20
Vermont	10	6	6	6	6	6
Virginia	25	16	12	17	11	9
West Virginia	8	5	4	6	4	4
Illinois	29	20	16	21	15	12
Indiana	39	27	21	28	18	14
Iowa	17	10	7	10	7	4
Kentucky	100	100	100	100	100	100
Michigan	36	25	20	26	18	14
Minnesota	11	7	4	7	4	3
Missouri	4	1	1	2	1	0
Ohio	46	35	28	36	25	19
West Virginia	100	100	100	100	100	100
Wisconsin	22	15	12	16	11	9
	32	22	17	23	16	12
California	0	0	0	0	0	0
Idaho	5	3	2	3	2	1
Montana	0	0	0	0	0	0
Nevada	0	0	0	0	0	0
Oregon	5	4	3	4	2	2
Washington	12	8	6	8	5	4

Source: Durfee and Coleman (1983), pp. 132-49.

been made. The results are summarised in Table 6.6. It is interesting that there is not much difference between the two subcategories of sites which suggests a broad consistency in siting practices over time.

A comparison of US and UK demographic siting criteria

At this point it is very useful to compare the demographic criteria

TABLE 6.6 *Effects of alternative US siting criteria on existing and proposed sites*

Category of sites	% of satisfying alternatives						
	1	2	3	4	5	6	All 6
Existing sites	53	70	76	68	79	83	52
Possible sites	58	68	76	65	78	84	58

Source: Durfee and Coleman (1983), pp. 185-9.

that have evolved in both the US and UK. A simple comparison is made difficult by the use of different sizes of radial zones and sectors. Moreover, the later US criteria are expressed as population density constraints. Other problems concern the fact that the 1962 US guidelines are incapable of a simple numeric representation except to the extent that the 1975 population density guidelines might be regarded as being what was intended by the 1962 criteria. In the UK the 1962 Farmer criteria were specified as neither densities nor population counts; however, for the special case of a uniformly distributed population it is possible to estimate those population densities which would conform to class I sites, using numerical methods. Finally, 2 of the 6 alternative 1984 siting criteria described by Kelly *et al.* (1984) are used to represent current US intentions; alternative 1 which is the most restrictive and alternative 6 which is the least restrictive.

Table 6.7 is an attempt to compare various US and UK demographic siting criteria in terms of their radial density and population limits. The results are fascinating. By far the least restrictive are the so called UK 'remote' siting criteria of 1962 which are a factor of two less restrictive than the subsequent relaxed criteria; however, this 'poor' performance of the Farmer (1962) criteria is critically dependent on the assumption of a uniform population distribution. It is also interesting to observe that the UK relaxed siting criteria are between 3 and 4 times more relaxed than contemporary US criteria.

Table 6.8 continues the comparison by examining equivalent density and population counts for a standard $22\frac{1}{2}$ degree sector. On this basis the Farmer (1962) criteria are now far more similar to the Marley and Fry (1955) remote siting criteria than previously, although the UK's relaxed criteria are now between 4 and 24 times more relaxed than any of the US criteria.

It must be emphasised that the simple horizontal comparisons

219

TABLE 6.7 *A comparison of various UK and US demographic siting criteria for radial populations*

Distance band (miles)	UK 1955 remote siting	1962 remote siting	1973 relaxed siting	USA 1975 density guidelines	1984 alternatives 1	6
Densities, persons per square mile						
0-0.3	low	low	0	500	250	250
0-1.5	424	424	2062	500	250	250
0-5	763	4558	2248	500	450	670
0-10	1910	4558	2134	500	490	730
0-20	none	4558	2005	500	497	747
0-30	none	none	none	500	499	748
Population limits						
0-0.3	low	low	0	141	71	71
0.3-1.5	3000	3000	14572	3393	1696	1696
1.5-5	57000	340764	161971	35876	34432	50926
5-10	543000	1077240	508572	121204	119506	178410
10-20	none	4637748	2011304	507115	505670	757784
20-30	none	none	none	906602	904903	1356505

TABLE 6.8 *A comparison of UK and US demographic siting criteria for $22\frac{1}{2}$ degree sectors*

Distance band (miles)	UK 1955 remote siting	1962 remote siting	1973 relaxed siting	USA 1975 density guidelines	1984 alternatives 1	6
Densities, persons per square mile						
0-0.3	low	low	0	500	250	250
0-1.5	5093	4558	10438	500	250	250
0-5	1528	4558	12576	500	920	2680
0-10	3820	4558	12766	500	979	2919
0-20	none	4558	12782	500	995	2980
0-30	none	none	none	500	997	2991
Population counts						
0-0.3	low	low	0	8	4	4
0.3-1.5	1125	375	4611	212	106	106
1.5-5	21375	21110	57123	2238	4410	13049
5-10	203625	67515	193542	7580	14832	44285
10-20	none	289671	810384	31689	63315	189763
20-30	none	none	none	56668	113006	338809

obtained from Tables 6.7 and 6.8 fail to indicate what may be termed vertical distribution effects. In practice these density and population counts provide a fairly poor indication of the numbers of acceptable sites that may be found because populations are not uniformly distributed. The relationship between the lumpiness and denseness of population distributions can significantly affect the behaviour of the siting criteria; as indeed does the range over which the criteria are applied. The same density criteria applied to 5, 10, 20, and 30 miles may yield very different results. Thus it may be concluded that it is the unknown interaction effects between denseness and lumpiness in population distributions, which reflect the historic patterns and scales of urbanisation, and the geographical limits over which they are applied that conditions the behaviour of the demographic criteria.

This apart, the marked differences in siting criteria also reflect more fundamental differences in safety philosophy. The obvious explanation that the differences in siting criteria merely reflect differences in the population density of the UK and US does not stand close examination. It should not be forgotten that parts of the US are as densely populated as the UK. For example, the Pennsylvania-New Jersey-Maryland Interconnection service area is densely populated and has a generation capacity similar to the whole of the UK and a total nuclear capacity also similar to that of the CEGB. The real reason for the apparent differences in siting practices is simply that the population centre distance rule has been applied in the US and this restriction, more than any other, has precluded development on sites which are in close proximity to urban concentrations. But this siting rule is itself a reflection of the traditional US emphasis given to reactor accidents beyond that which the design can handle; in the UK only design basis accidents have been considered. So the major differences in siting practices which are reflected in Tables 6.7 and 6.8 reflect this major difference in safety philosophy.

Application of US demographic criteria to the UK

A most interesting exercise is to apply the various US siting guidelines to the UK. This is relevant both as a means of examining the actual effects of the previously identified differences in siting criteria and because of current proposals to build US designed reactors in the UK and to site them according to the existing UK relaxed siting criteria. The prospect may well arise whereby PWRs could be built on sites in the UK which would be regarded as totally unacceptable in the US. The obvious CEGB response would probably be to point out the

221

greatly enhanced safety characteristics of the British built, British operated, and British managed PWR; although whether any of these factors confers any great improvement in safety is a matter for debate, and totally untested when the decisions to build are made.

The early US siting criteria are difficult to apply to UK sites because they were only used as guidelines and were interpreted in a flexible way. Engineering improvements have been used to compensate for site deficiencies so it is not possible to estimate precisely what distance factors would have been applied to UK reactors. On the other hand, the AEC's numerical criteria (1963) were meant to reflect current practice at that time and they can be applied, sensu strictu, to UK sites by using the AEC equations to estimate the radii of the three critical zones as functions of reactor sizes which are located there. A guide as to the likely acceptability or otherwise of the UK sites can be obtained by computing the 1971 populations within these three notional areas assuming that no credits have been given for engineering design measures (Openshaw, 1984).

The results in Table 6.9 indicate that very few of the UK sites would probably satisfy the population centre distance restrictions; principally Dounreay, Winfrith and Trawsfynydd. The low population zone populations are also generally far too large and at two sites (Sizewell and Druridge) the populations within the exclusion zones would be regarded as excessive. It should also be noted that the UK reactors generally do not have secondary containment systems as commonly found in the US. This would probably have resulted in very little credit being given to engineered safety measures despite the greater intrinsic safety of gas-cooled reactors. The main way in which US reactor designers allowed the distance factors to be reduced was by designing highly effective containment structures.

The situation regarding the 1975 population density guidelines as operationalised in the form of a 30-mile site population factor is a little more definite. An SPF greater than 0.500, or 500 persons per square mile, would be used to trigger a more detailed evaluation of alternative sites. On this basis, Table 6.9 suggests that Berkeley, Oldbury, Hartlepool, Heysham, and Druridge might well have been rejected unless no satisfactory alternative sites could be found within the relevant CEGB regions; and really, it is most unlikely that no better alternatives could be found.

The situation regarding the latest proposed demographic criteria is, if anything, even worse. Table 6.10 shows that very few sites would satisfy all six alternatives; only Dounreay,

TABLE 6.9 *Populations within critical USAEC distances and SPF's for UK power station sites in 1981*

Site	1981 populations (in '000s) population exclusion zone[a]	low population zone[a]	Population centre distance[a]	30-mile site population factor
UKAEA				
Chapelcross	.0	13.5	15.0	.15
Dounreay	.0	1.9	11.2	.01
Harwell	.0	24.7	47.2	.57
Windscale	.0	6.5	53.9	.11
Winfrith	.0	8.9	13.0	.25
Remote sites				
Berkeley	.0	49.8	93.7	.58
Bradwell	.0	14.3	57.3	.44
Dungeness	.0	45.4	208.5	.15
Hinkley Point	.0	74.2	172.6	.37
Hunterston	.0	132.8	367.1	.42
Oldbury	.0	122.7	419.3	.67
Sizewell[b]	11.2	338.6	605.2	.16
Trawsfynydd	.0	16.8	22.9	.08
Wylfa	.4	41.5	51.2	.06
Relaxed sites				
Hartlepool	.0	664.3	914.0	1.11
Heysham	.8	277.0	510.4	.55
Torness	.0	18.5	40.7	.08
Druridge Bay[b]	47.7	1218.3	1636.7	.51

Notes: a distances based on largest existing or proposed reactor.
b assumes PWR.

Dungeness, Trawsfynydd, and Torness.

An obvious response is to resort to the old argument that suitable sites are difficult to find in such a densely populated country as Britain. Once again it is argued that this assumption is fundamentally incorrect if suitability is defined in demographic rather than in terms of optimising the CEGB's engineering requirements. Table 6.11 gives an indication of the land area of the UK that would satisfy the 1975 population density guideline of 500 persons per square mile, as well as for a series of other density values. It would seem that large proportions of both the coastal areas and the total land surface would satisfy all but the

223

TABLE 6.10 *Effects of alternative US siting criteria on the acceptability of current UK reactor sites*

Site	US alternative criteria 1	2	3	4	5	6
UKAEA						
Chapelcross	fails	fails	fails	fails	fails	fails
Dounreay						
Harwell	fails	fails	fails	fails	fails	fails
Windscale	fails*	fails*	fails*	fails*	fails*	fails*
Winfrith	fails	fails	fails	fails	fails	fails
Remote sites						
Berkeley	fails	fails	fails	fails	fails	fails
Bradwell	fails	fails	fails	fails*		
Dungeness						
Hinkley Point	fails	fails	fails	fails*		
Hunterston	fails	fails	fails	fails*	fails*	fails*
Oldbury	fails	fails	fails	fails	fails	fails
Sizewell	fails	fails	fails	fails	fails	fails
Trawsfynydd						
Wylfa	fails*	fails*		fails*	fails*	
Relaxed sites						
Hartlepool	fails	fails	fails	fails	fails	fails
Heysham	fails	fails	fails	fails	fails	fails
Torness						
Druridge Bay	fails	fails	fails	fails	fails	fails

Note: * Fails sector limit only.

most extreme population density restrictions. The effects of using 20, 40, and 50 mile summations are minor. It would seem that the 1975 US density restrictions are only slightly more restrictive than the UK's own relaxed siting criteria. In practice these results would imply that there are many potential nuclear power station sites and there really ought to be a more rigorous and systematic way of distinguishing between 'good' and 'bad' reactor sites.

Figures 6.2 and 6.3 show those areas where 1km grid-squares fail the 500 and 1000 persons per square mile density constraints. These results can also be expressed as proportions of both the coastal areas and total are of each County; see Appendix 3, Table 1.

The proposed new US demographic criteria can also be assessed for their impact on the UK. Table 6.12 shows the areas that would be excluded by the various criteria. It would appear

TABLE 6.11 *Effects of 1975 US density guidelines on the availability of UK sites*

30 mile SPF density limit[a]	Percentage of area excluded coastal areas	total land
100	40.6	58.1
200	30.0	43.3
300	24.0	32.8
400	18.8	25.8
500*	15.5	21.2
600	12.8	18.0
700	10.8	15.5
800	8.7	13.4
900	7.5	11.3
1000**	6.3	10.5
1100	5.4	9.2
1200	4.6	8.2
1300	3.9	7.3
1400	3.4	6.5
1500	3.0	5.9

Notes: * density limit of 500 persons per square mile used to trigger an in-depth review of alternative sites.
** density limit of 1000 persons per square mile used as a life-time limit on reactor sites.
a persons per square mile.

TABLE 6.12 *Effects of alternative US siting criteria on the availability of UK sites*

US criteria	Percentage of area excluded coastal areas	total land
1	44.0	46.8
2	38.7	41.3
3	35.5	38.0
4	41.7	44.2
5	35.9	38.3
6	32.7	35.1

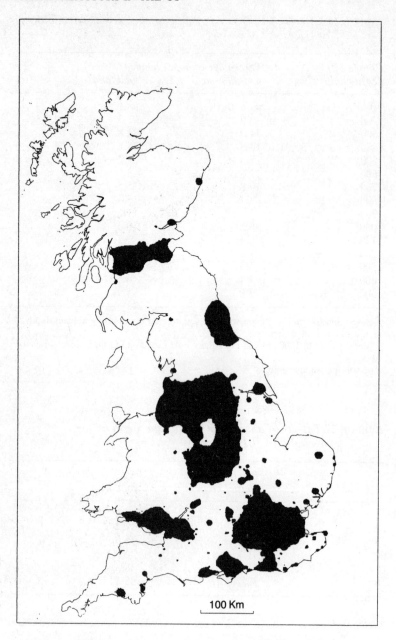

FIGURE 6.2 Areas in the UK with average 30-mile population densities greater than 500 persons per square mile.

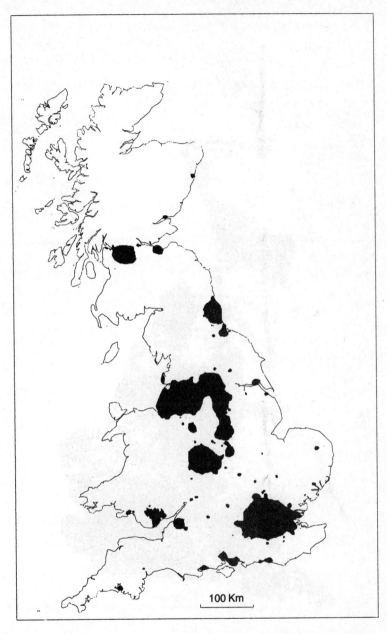

FIGURE 6.3 Areas in the UK with average 30-mile population densities greater than 1000 persons per square mile.

FIGURE 6.4 *Areas in the UK which fail US siting criteria alternative 1.*

FIGURE 6.5 *Areas in the UK which fail US siting criteria alternative 2.*

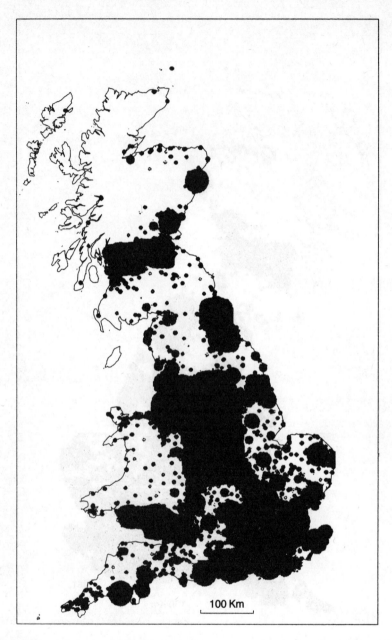

FIGURE 6.6 *Areas in the UK which fail US siting criteria alternative 3.*

FIGURE 6.7 *Areas in the UK which fail US siting criteria alternative 4.*

FIGURE 6.8 Areas in the UK which fail US siting criteria alternative 5.

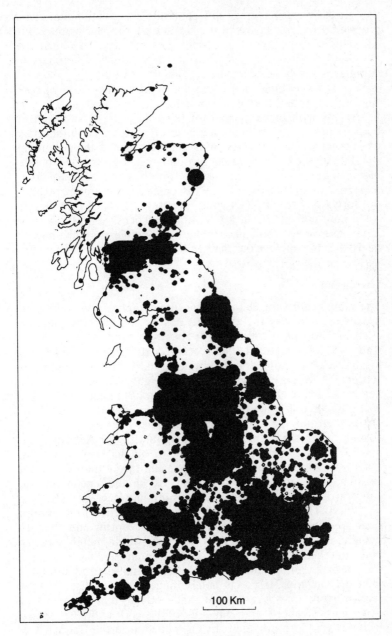

FIGURE 6.9 Areas in the UK which fail US siting criteria alternative 6.

that these exclusion areas are generally no greater than those reported for various parts of the North-eastern US. Figures 6.4 to 6.9 show the location of infeasible sites for these six alternative US demographic criteria. A more detailed statistical summary by counties is given in Appendix 3, Tables 2 and 3.

The overwhelming impression gained from these comparative site impact studies is that the situation in the UK compares very favourably with that reported for the US. There would seem to be no a priori basis for continuing to argue for a major shortage of suitable reactor sites. It is most unlikely that *all* the potentially suitable sites on demographic grounds can be excluded on the basis of other criteria. In the US this has not been found to happen although in the UK this is precisely the argument often put forward by the CEGB and, indeed, the NII. Clearly there is something sadly amiss with UK siting practices. Suggestions as to how a more reasonably and publicly acceptable approach may be devised is the subject of Chapter 8.

Conclusions

The main reason for changing the demographic siting criteria in use in the US was the desire to reduce the risks of radiation exposure to the surrounding population. Indeed two recent studies have suggested that certain existing sites which are located near to population centres should be recommended for retirement because of the relatively high population densities nearby (Burwell and Lane, 1980; Cope and Baumann, 1977). It is also possible that few new sites will be needed in the future if full use is made of the expansion potential at existing sites; this is thought to be sufficient to last at least until 2025 AD (Briggs *et al.*, 1978). A critical factor appears to be that any new siting criteria should not rule out the expansion of the majority of the existing sites; and that, moreover, it should encourage the concept of cluster siting at both existing and new sites (Kelly *et al.*, 1984). The main reasons are that the majority of the existing sites are already remote from population centres and that the opening up of large numbers of new sites would merely increase both the land area and the population at significant risk of exposure. Other advantages of cluster over dispersed siting are seen to include: (i) a broader range of expertise; (ii) an improvement in performance and safety that comes with experience (Roberts and Burwell, 1981); (iii) additional special facilities could be provided; (iv) a reduction in the frequency of accidents which more than offsets the increased risks from an accumulation of reactors; (v) a reduction in offsite exposure because the site area is larger and the total number of sites and thus the regions of potential danger are reduced; and (vi) a favourable community reaction because most of the plant employees will live locally.

Burwell and Lane (1980) speculate that some new sites will still be required. They suggest that of the 91 existing sites, 19 might be eventually retired as unsuitable for expansion because of population density and land development constraints, and the remaining sites augmented by 29 new or not yet opened sites by 2025 AD. They also indicate that a population density threshold of 300 persons per mile out to 10 miles is an effective discriminant between sites with significantly higher nearby population densities from those with near average or low densities.

The question arises, however, as to whether Kelly et al. (1984) are being realistic. It seems that the original objective of the SPTF to re-establish siting as a safety measure has been replaced by a concern to find new siting criteria that render most of the existing sites acceptable. Yet the implications of the SPTF's recommendations are that changes are needed precisely because it seemed that most existing sites had been selected with insufficient attention being given to geographical isolation as a safety measure. An indication of the apparent relaxation since 1979 can be obtained by comparing the population levels allowed in Table 6.3 with the equivalent figures for the illustrative densities suggested in NRC (1979). The latter values implied cumulative radial population limits of 7775 persons for 0 to 5 miles, 43117 with 10 miles, and 420108 within 20 miles; although they could have been increased by the use of regional population densities. These illustrative values are considerably lower than the equivalent cumulative values of 36128 or 52622 for 0 to 5 miles, 153938 or 229336 for 0 to 10 miles, and 625176 or 936194 for 0 to 20 miles for the 6 alternative criteria in Table 6.3. On the other hand the Table 6.3 values are consistent with Regulatory Guide 4.7 (1975b) which specified a notional density of 500 persons per square mile out to 30 miles and a lifetime limit of 1000 persons per square mile; as Table 6.2 shows the 6 alternative demographic criteria are based on densities of 500 to 750 persons per square mile out to 30 miles. So it appears that the criteria being examined in 1984 differ only slightly from those suggested in 1975; the main difference being the use of sector limits as well as radial limits.

It appears that siting is once again being affected more by a concern about site availability than public safety considerations. The various demographic analyses that have been performed would seem to indicate that a more stringent set of siting criteria could be devised that would still leave potential power reactor sites in all regions of the US. However, this constraint may itself be irrelevant. If the future of nuclear power is seen to involve fast breeders and integrated fuel processing and reprocessing facilities then national rather than regional siting will probably be required. It is slightly surprising the US should seem to be preparing only for a thermal reactor future when proximity to regional load centres may still retain some significance. It is also surprising that the apparently high level of public concern about nuclear safety in the US, and the large-scale geographical impact of Three Mile Island, should not have resulted in a far more radical set of siting criteria proposals. The origins for such developments are present in the report of the SPTF, but it remains to see to what extent they materialise in practice, even if it looks as if it may be a largely academic matter for the rest of this century.

Appendix 3

TABLE 1 *Areas of UK counties that exceed 500 and 1000 persons per square mile density constraints*

Region County	I	II	III	IV
Scotland				
Borders	0	0	0	0
Central	100	0	33	4
Dumfries & Galloway	0	0	0	0
Fife	58	4	41	1
Grampian	5	1	1	0
Highland	0	0	0	0
Lothian	27	18	31	8
Strathclyde	4	1	16	8
Tayside	19	3	2	0
Islands	0	0	0	0
East Anglia				
Cambridgeshire	100	100	4	0
Norfolk	5	0	3	0
Suffolk	10	0	3	0
East Midlands				
Derbyshire	100	100	68	26
Leicestershire	100	100	64	20
Lincolnshire	0	0	1	0
Northamptonshire	100	100	18	1
Nottinghamshire	100	100	73	37

TABLE 1 *cont.*

North				
Cleveland	78	41	77	35
Cumbria	4	0	0	0
Durham	100	86	39	13
Northumberland	20	13	9	3
Tyne & Wear	100	100	100	96
North West				
Cheshire	100	98	71	34
Greater Manchester	100	100	100	100
Lancashire	71	14	70	45
Merseyside	98	85	100	100
South East				
Bedfordshire	100	100	61	13
Berkshire	100	100	55	36
Buckinghamshire	100	100	47	21
East Sussex	45	9	26	4
Essex	45	25	49	31
London	100	100	100	100
Hampshire	90	58	41	13
Hertfordshire	100	100	89	66
Isle of Wight	28	0	19	0
Kent	59	12	40	18
Oxfordshire	100	100	21	2
Surrey	100	100	87	57
West Sussex	68	11	46	4
South West				
Avon	85	18	90	29
Cornwall	5	1	1	0
Devon	16	4	4	1
Dorset	26	5	11	3
Gloucestershire	13	0	12	1
Somerset	1	0	1	0
Wiltshire	100	100	8	0
West Midlands				
Hereford & Worcester	100	100	13	3
Salop	100	100	8	0
Staffordshire	100	100	77	37
Warwickshire	100	100	52	29
West Midlands	100	100	100	100
Yorkshire & Humberside				
Humberside	24	8	13	3
North Yorkshire	5	0	13	2
South Yorkshire	100	100	95	61
West Yorkshire	100	100	100	91
Wales				
Clwyd	48	10	11	0
Dyfed	2	0	1	0
Gwent	100	9	58	21

TABLE 1

Region County	I	II	III	IV
Gwynedd	0	0	0	0
Mid Glamorgan	4	0	89	33
Powys	100	100	1	0
South Glamorgan	67	35	67	32

Notes: I coastal areas with densities greater than 500 persons per square mile.
II coastal areas with densities greater than 1000 persons per square mile.
III total land area with population densities greater than 500.
IV total land area with population densities greater than 1000.
All results based on calculating site population factors for a 30 mile radius.
Counties which are landlocked have all their coastal area excluded. The 1975 US reactor siting guidelines would give preference to sites in areas with population densities of less than 500 persons per square mile and generally prohibit siting in areas where densities exceed 1000 persons per square mile. According to these criteria, there is no shortage of sites in the UK away from the metropolitan counties.

TABLE 2 *Coastal areas of UK counties excluded by the six US revised siting suggestions*

Region County	I	II	III	IV	V	VI
Scotland						
Borders	65	50	50	50	50	50
Central	100	100	100	100	100	100
Dumfries & Galloway	18	11	9	15	10	8
Fife	96	83	82	90	82	81
Grampian	65	52	42	60	46	38
Highland	8	5	4	7	5	4
Lothian	59	59	59	59	59	56
Strathclyde	19	14	12	16	12	11
Tayside	96	89	86	90	84	77
Islands	2	1	1	2	1	1
East Anglia						
Cambridgeshire	100	100	100	100	100	100
Norfolk	79	68	62	69	66	53
Suffolk	85	80	70	81	70	67
East Midlands						
Derbyshire	100	100	100	100	100	100
Leicestershire	100	100	100	100	100	100

TABLE 2 *cont.*

Lincolnshire	41	22	15	27	14	10
Northamptonshire	100	100	100	100	100	100
Nottinghamshire	100	100	100	100	100	100
North						
Cleveland	100	100	95	100	100	95
Cumbria	80	72	57	74	58	49
Durham	100	100	100	100	100	100
Northumberland	58	50	46	54	50	45
Tyne & Wear	100	100	100	100	100	100
North West						
Cheshire	100	100	100	100	100	100
Greater Manchester	100	100	100	100	100	100
Lancashire	100	100	100	100	100	91
Merseyside	100	100	100	100	100	100
South East						
Bedfordshire	100	100	100	100	100	100
Berkshire	100	100	100	100	100	100
Buckinghamshire	100	100	100	100	100	100
East Sussex	97	93	91	93	87	86
Essex	100	100	96	100	90	85
London	100	100	100	100	100	100
Hampshire	100	100	100	100	100	100
Hertfordshire	100	100	100	100	100	100
Isle of Wight	100	97	85	100	76	74
Kent	96	95	95	96	92	90
Oxfordshire	100	100	100	100	100	100
Surrey	100	100	100	100	100	100
West Sussex	100	100	100	100	100	100
South West						
Avon	100	100	100	100	100	100
Cornwall	70	61	54	68	55	48
Devon	86	80	72	83	74	68
Dorset	100	89	84	92	82	77
Gloucestershire	100	92	90	87	76	57
Somerset	84	65	62	72	62	57
Wiltshire	100	100	100	100	100	100
West Midlands						
Hereford & Worcester	100	100	100	100	100	100
Salop	100	100	100	100	100	100
Staffordshire	100	100	100	100	100	100
Warwickshire	100	100	100	100	100	100
West Midlands	100	100	100	100	100	100
Yorkshire & Humberside						
Humberside	100	100	93	100	91	79
North Yorkshire	91	67	60	76	58	45
South Yorkshire	100	100	100	100	100	100
West Yorkshire	100	100	100	100	100	100

239

TABLE 2 *cont.*

Region County	I	II	III	IV	V	VI
Wales						
Glwyd	100	100	100	100	100	100
Dyfed	50	31	22	42	27	22
Gwent	100	100	100	100	100	100
Gwynedd	59	42	35	53	41	35
Mid Glamorgan	100	81	77	86	77	63
Powys	100	100	100	100	100	100
South Glamorgan	100	100	98	100	94	94

Notes: I, II, III, IV, V, and VI refer to US siting alternatives 1 to 6 as discussed in Chapter 6.

TABLE 3 *Total areas of UK counties excluded by the six US revised siting suggestions*

Region County	I	II	III	IV	V	VI
Scotland						
Borders	23	16	12	19	13	10
Central	58	51	49	52	48	45
Dumfries & Galloway	16	12	9	14	11	8
Fife	98	87	82	90	78	71
Grampian	29	20	16	24	17	11
Highland	4	2	1	3	2	1
Lothian	76	73	70	74	69	63
Strathclyde	37	32	29	33	29	26
Tayside	41	29	25	31	23	19
Islands	3	2	1	3	2	1
East Anglia						
Cambridgeshire	96	78	69	83	61	46
Norfolk	71	55	49	64	43	37
Suffolk	77	64	50	69	52	42
East Midlands						
Derbyshire	100	100	100	100	94	88
Leicestershire	100	96	95	97	95	90
Lincolnshire	72	51	38	58	38	28
Northamptonshire	98	83	77	86	71	65
Nottinghamshire	100	98	98	100	93	90

TABLE 3 *cont.*

North						
Cleveland	100	100	98	100	97	92
Cumbria	42	31	24	34	23	19
Durham	67	59	56	60	56	51
Northumberland	35	27	24	30	24	21
Tyne & Wear	100	100	100	100	100	100
North West						
Cheshire	100	95	93	94	85	82
Greater Manchester	100	100	100	100	100	100
Lancashire	100	95	93	95	89	84
Merseyside	100	100	100	100	100	100
South East						
Bedfordshire	100	100	100	100	97	85
Berkshire	96	88	83	88	84	76
Buckinghamshire	98	91	89	94	81	77
East Sussex	100	100	97	97	81	75
Essex	97	92	89	93	86	79
London	100	100	100	100	100	100
Hampshire	98	95	90	95	84	78
Hertfordshire	100	100	100	100	95	94
Isle of Wight	100	96	84	100	78	58
Kent	97	91	88	91	82	78
Oxfordshire	98	89	79	93	78	70
Surrey	100	100	100	100	97	96
West Sussex	100	100	100	95	86	80
South West						
Avon	100	100	100	100	100	100
Cornwall	66	53	44	61	45	38
Devon	63	50	43	54	42	32
Dorset	80	62	53	67	53	44
Gloucestershire	96	86	77	89	73	59
Somerset	74	51	45	60	47	41
Wiltshire	83	68	57	72	56	48
West Midlands						
Hereford & Worcester	76	54	50	57	44	38
Salop	60	47	36	52	36	29
Staffordshire	100	100	100	100	98	97
Warwickshire	92	82	79	85	77	73
West Midlands	100	100	100	100	100	100
Yorkshire & Humberside						
Humberside	95	89	81	90	74	64
North Yorkshire	65	53	42	57	42	35
South Yorkshire	100	100	100	100	100	98
West Yorkshire	100	100	100	100	100	100
Wales						
Clwyd	58	44	38	47	35	31
Dyfed	34	23	16	29	18	16

241

TABLE 3 *cont.*

Region County	I	II	III	IV	V	VI
Gwent	100	88	83	89	76	71
Gwynedd	32	20	17	29	19	16
Mid Glamorgan	100	98	98	100	97	96
Powys	24	16	12	19	11	9
South Glamorgan	100	100	98	100	95	93

7 Demographic characteristics of nuclear power station sites in the UK and US

At this point it is useful to consider the provision of some basic demographic statistics for current nuclear power station sites in the UK and the US for comparison purposes. Such data might well prove useful to those interested in the siting of nuclear power stations and may be used to influence future policies. It could also be used as background briefing material for persons, in or outside of government, in the siting aspects of nuclear power plant. For example, what are the basic demographic characteristics of the various sites? How do they compare with each other? How do UK sites compare with US sites? Can any distinctive types of site be recognised by their demographic characteristics. Some of the related, more specific questions concerning the evaluation of siting policies and practices have already been answered in the preceding chapters but these more general ones remain to be briefly answered here. However, before tackling these aspects there are a few outstanding methodological problems concerning the process of providing demographic data for particular sites which have to be considered first. These problems apply to all the demographic analyses reported in this book, and they are not without a considerable degree of significance because of their implications for the safety assessment studies of nuclear power stations.

Some technical problems in providing estimates of demographic data for nuclear sites

The basic problem is easily defined. The objective is to obtain population estimates for a series of radial zones divided into between 12 and 36 angular sectors. This spatial data retrieval pattern is common to all the siting criteria examined in Chapters 3, 4, and 6 and the various zoning systems embodied in these

243

criteria are summarised in Figure 7.1a, b, c, d. The data which are to be aggregated only exist in the UK for either irregularly shaped census enumeration districts or for geometrically regular 1km grid squares. In the US only census enumeration areas are available for use. The problem in both countries is the same. The

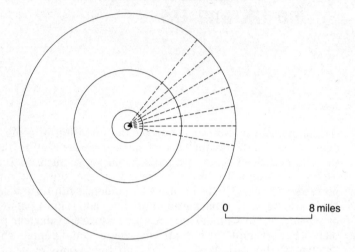

0 _____ 8 miles

FIGURE 7.1a Sector and annular zones used by Marley and Fry's remote siting criteria (1955)

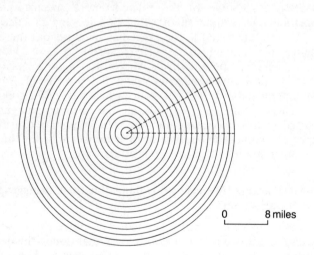

0 _____ 8 miles

FIGURE 7.1b Sector and annular zones used by Farmer's remote siting criteria (1962)

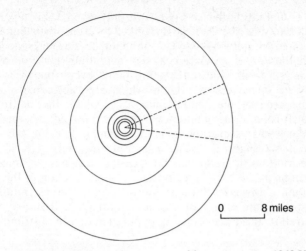

FIGURE 7.1c Sector and annular zones used by various post-1968 UK
relaxed siting criteria

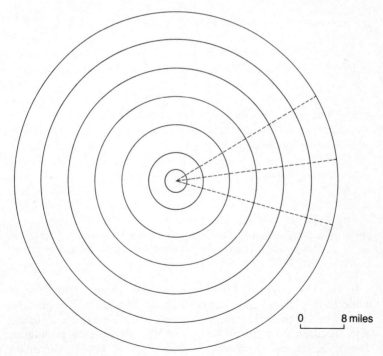

FIGURE 7.1d Sector and annular zones used by proposed new US siting
criteria

245

census enumeration districts (eds) are only given a single point of reference to describe their geographical location; their boundary coordinates are not available yet for complete national spaces. As a result the entire population of the enumeration district has to be assumed to be located at this point even though it may actually be dispersed over areas which are substantially larger than the area of the spatial interval in which the centroid is located. It is possible then the centroids of those eds assigned to a particular radial sector or zone may contain people who live outside it; see Figure 7.2. The results of these 'edge effects' may well be random for radii beyond a certain distance, because the populations in the border regions are a small fraction of the total population. However, little is known about the magnitude of these estimation errors or of the interactions between zonal configuration and the underlying population density distributions.

FIGURE 7.2 *An illustration of the edge effect problem when irregularly shaped census areas are assigned by their centroids to segments of a circle*

In the US a typical census ed contains about 1,000 persons, while in the UK a range between about 20 and 700 persons may be typical so there is plenty of scope for edge effects to occur. The importance of this problem probably decreases as population density increases. According to Aldrich *et al.* (1982), the errors are likely to be negligible beyond 20-mile distances even for sparsely populated regions, while beyond 7 miles the errors are unlikely to be substantial for population densities greater than 500 persons per square mile. In the UK these errors might well be less because the eds are smaller and the country has high average population densities; nevertheless, the magnitude of these estimation errors may still be sufficient to question the results obtained from census based site assessments.

Another problem in the UK is that national census population data for eds are only readily available for 1981; royalty charges on the 1971 data have prevented their use at the national level. The only available 1971 population data relates to 1km grid-squares (Rhind et al., 1977) and these data have been widely used in nuclear safety studies (Hallam et al., 1980). The relatively poor level of geographical resolution (1km compared with the 100 metre references on the ed data) can be overcome by assuming that the population density in each 1km square is uniform and then computing the fraction of each 1km square located within the various radial and sectoral zones. This should reduce the occurrence of possibly large edge effects but it will also tend to smooth or average out localised population concentrations. Nevertheless, if 1km grid-square counts are to be used then this would seem to be a workable solution.

In the US the initial simple centroid assignment techniques used by Bunch et al. (1979), Robinson and Hansen (1981), and Aldrich et al. (1982) have been replaced by more sophisticated distance weighted interpolation and normalisation procedures; see Durfee and Coleman (1983). They estimate that 'At the county level some distributions were improved by as much as 30 per cent or so with this new normalization' (p. 48). However, there are problems with these interpolation methods, particularly their tendency to assume continuous population distributions even in areas of low or zero populations. Further improvements may be made if detailed digital outlines are available for various zonal features and when it is possible to integrate remote sensing data with census material. However, none of these exciting future developments really help with the current problems of obtaining reasonably accurate population estimates for regions around existing nuclear sites. The more sophisticated US interpolation methods cannot yet be applied in the UK because census ed areas are not known and these are needed for the normalisation techniques that have been developed.

Other problems with census data concern the fact that they describe a night-time population distribution which may offer little or no guidance about daytime distributions either at weekends or during normal working days. Quite simply the extent to which populations may vary over short periods of time

247

is not known for small census areas. Some guidance for larger units, for example census wards in the UK, can be obtained from the journey to work tabulations, but these data fail to provide any reflection of shopping, recreation, or school trips and thus offer no real guidance as to likely daytime population distributions. All that can be said is that given the existing lack of data for daytime distributions it is possible, even likely, that the temporal fluctuations in population numbers during the day will totally swamp any errors due to technical problems in estimating accurate night-time distributions.

Estimating the magnitude of population errors

These problems apart it is still relevant to try and identify the likely magnitude of the estimation errors contained in typical demographic data for nuclear sites. Without access to a complete set of geographically referenced individual census data all that can be done is to conduct a few empirical experiments. The author was fortunate in having access to a large land-use data base for Tyne & Wear County in north-east England, UK. The location of all houses, basically households, have been digitised to a resolution of 1 metre. In this area of 1.1 million people there is a set of 448,808 household references. However, no information is available about household sizes, so the simple assumption is made that each is of equal importance for the purposes of the present experiments. A series of arbitrary locations are selected within this area so that the number of houses within a set of 7 distance bands and 12 30 degree sectors could be estimated using a centroid assignment technique. The distance bands are those used for the relaxed UK siting criteria since this provides an extreme test due to the narrow width of the first four distance bands. It is assumed that the 1 metre referenced data can be accurately assigned by their centroids.

Four sites are examined; a periurban coastal site typical of a UK relaxed site, a periurban site located halfway between two major urban areas, a site located in the midst of the region on a river, and an existing coalfired power station.

The 1 metre referenced data was aggregated to form a set of 549 1km grid-squares. It was not possible to use census eds here because no coordinate boundary data were available. The 1km data were then assigned to the same distance bands and sectors as the original unaggregated data using an area based prorated estimation technique. The results for the seven distance bands are shown in Table 7.1.

As might be expected the largest relative errors occur in the 0-

TABLE 7.1 *Errors in estimating populations within radial distance bands*

Distance band (miles)		Periurban coastal site	Periurban midway site	Central urban site	Existing coal fired power station site
0-1	actual	705	2099	5925	4758
	estimated	838	1838	6431	4332
	% error	−19.2	12.4	−8.5	8.9
1-1½	actual	1335	498	16551	8989
	estimated	1351	817	17199	9570
	% error	−1.2	−64.1	−3.9	−6.4
1½-2	actual	1987	2218	24224	13470
	estimated	2116	3145	22799	11971
	% error	−6.5	−41.8	5.8	11.1
2-3	actual	10522	20479	44002	27499
	estimated	10456	19629	43899	27325
	% error	.6	4.1	.2	.6
3-5	actual	47211	67052	91738	53755
	estimated	46432	66960	90781	55903
	% error	1.6	.1	1.0	−3.9
5-10	actual	111080	210734	233848	158852
	estimated	111054	208666	234890	158237
	% error	.0	.9	−.4	.3
10-20	actual	275946	145728	32521	181486
	estimated	276531	147750	32807	181468
	% error	−.2	−1.3	−.8	.0

to 2-mile distance bands with the size of the errors varying according to the characteristics of the site being examined. The 1km prorated estimates tend to exaggerate the close-in populations and may be regarded as conservative; a useful property in this business. Beyond 3 miles the errors become very small and generally insignificant. If the 1km squares had been assigned using a simple centroid assignment with the whole cell being allocated according to whatever zone its centroid lay within, then somewhat larger errors would have been produced. For example, for the periurban site errors of 312, −37, −1046, 442, 3020, 5364, and −8055 would be produced; percentage errors of 14.8, −7.4, −47.1, 2.1, 4.5, 2.5, and −5.5 respectively for the seven distance bands. Apart from the 1½- to 2-mile distance band these would not be particularly excessive and the close-in errors are generally smaller than the area prorated estimates.

TABLE 7.2 *Errors in estimating populations within 30 degree sectors*

	Sector		Distance bands 0-1	1-1½	1½-2	2-3	3-5	5-10	10-20
d	4	a	0	0		0	0	10	0
		b	4	2		27	48	948	0
		c	0	0		0	0	182	0
	5	a	53	5	919	4723	10076	41766	6254
		b	107	178	813	4688	10708	40481	6128
		c	239	5	802	3328	10106	41688	6033
	6	a	451	0	512	5002	17419	71761	94820
		b	367	16	655	4697	16955	70964	96841
		c	0	2	435	5552	16479	67779	96543
	7	a	699	14	291	1758	8756	47323	38527
		b	776	79	373	1652	9118	46801	38294
		c	1337	3	530	1423	10157	45787	43867
	8	a	815	321	130	3355	14725	27616	1248
		b	423	334	697	3637	14394	27681	1478
		c	0	525	955	3831	13224	27192	2294
	9	a	81	158	366	5639	16076	22258	4879
		b	82	203	607	4934	15720	21493	5009
		c	211	0	542	5903	14064	22742	5046
	10	a	0	0			0	0	
		b	48	6			17	299	
		c	0	0			0	0	
d	11	a	0						
		b	14						
		c	0						

a observed house counts
b estimated using area prorated 1km grid squares
c estimated using centroid assignment of 1km squares
d remaining 30 degree sectors have zero populations of houses
Note: sectors 1, 2, 3 and 12 have zero houses and are not shown.

Table 7.2 shows the errors for this periurban site for each 30 degree sector. Obviously the errors are larger for these smaller areas. Nevertheless, the 1km area prorated estimates have negligible errors for distances beyond 3 miles and occasionally may result in large overestimates within this area. The centroid assignment based estimates are poorer and the magnitude of these errors decreases more slowly with distance. The explanation is that the areas of the 30 degree sector segments of the closein distance bands are very small – 0 to 1 mile the area is 0.67 km², 1 to 1½ miles it is 0.84 km², 1½ to 2 miles it is 1.18 km², 2 to 3 miles it is 3.38 km² – in relation to the size of the census data units. The greater error propensity for the simple centroid assignments that are used in the automated site searches of the previous chapters is of little real consequence because the errors are largest in sparsely populated areas where they do not affect the performance of the siting criteria being applied. In any case the purpose of these automated site searches is strategic; that is to give a broad indication of the areas excluded by various demographic criteria. It would be possible to modify the computer algorithms to use an area prorated estimation technique but the computational costs would be between 5 and 10 times higher; the additional degree of precision was not worthwhile. This decision is reversed whenever particular sites are being examined.

It is thought reasonable that the errors involved with the 1981 census ed population counts will be less than those reported in Tables 7.1 and 7.2 although there is no empirical verification of this assumption.

It is also clear that the detailed study of particular locations requires data which have a higher degree of geographical resolution than that which is readily available from the census. This applies mainly when detailed evaluations are to be made of a small number of alternative sites. The CEGB use factored-up house counts obtained from various sources including field survey, aerial photography, and electoral registers. However, unless the data are computerised it is unlikely that anything other than fixed sectors can be applied. In any case the question arises as to whether data acquired in this way would be very different from that obtained from the census particularly as there are errors associated with all the census alternatives. At present the only way of assessing how different the population estimates may be is to compare a set of '1981 and beyond' CEGB population estimates for Hartlepool and Heysham with those obtained from census data sources. These CEGB estimates are only available for complete distance bands and they include an attempt to forecast changes during the lifetime of the nuclear stations involved; see Table 7.3.

The differences are large and it is difficult to know which set of figures are the most accurate. The CEGB estimates are believed to have been produced in the mid 1960s and to be based on a number of sources; house counts for the close-in areas and various planning forecasts for areas further away. They may not have been based on 1971 census small area data. This may explain why the CEGB estimates tend to over-estimate the populations beyond 10 miles because they were probably not as accurate as would be possible today and they might well have been based on large units, such as Local Authorities. The closein

251

TABLE 7.3 *Comparison of census population estimates with CEGB projections for Hartlepool and Heysham*

Distance band (miles)	Hartlepool site 1971 1km	1981 eds	CEGB	Heysham 1971 1km	1981 eds	CEGB
0-1	228	369	20	1457	2075	3500
1-1½	1794	2101	300	3446	3600	6000
1½-2	5574	1656	5000	3374	2853	7000
2-3	26801	22361	21800	11792	11605	45600
3-5	209042	138457	168500	45504	50698	70000
5-10	319161	278177	414000	91206	81002	140000
10-20	506874	511806	535000	476359	476165	500000
total	1069474	954927	1144620	633138	627998	772100

populations tend to be underestimates of the 1981 derived figures. For Hartlepool the differences in the 0 to 1½ mile populations would seem to far exceed the likely error estimates obtained from the numerical experiments in Table 7.1. The differences may well be attributable to real errors in the CEGB estimates. A similar tentative conclusion could also be applied to Heysham, particularly to the 1- to 3-mile distance bands. The differences in the 5- to 10-mile figures probably reflect CEGB allowances for the development of a new town which has not in fact materialised. It might well be speculated that the differences in sector values would be even greater. It should be noted that the difficulties of obtaining accurate night-time and day-time population counts around nuclear facilities provide grounds for considerable public concern. The task is not impossible and clearly additional resources should be allocated to it.

A selection of demographic statistics

The limited objective here is to present a broad range of useful demographic statistics as a guide to the characteristics of the various sites. Readers are invited to draw their own conclusions although some guidance is also given.

Appendix 4, (pp. 260-8) presents a collection of basic demographic statistics for current UK and US reactor sites. The immediate and very striking conclusion is that the UK sites, with one or two exceptions, are located in areas of far higher population density than found in the US. The reasons for this are not entirely, if at all, attributable to major differences in population density but reflect fundamental differences in both safety philosophy and in siting practices. These points have already been elaborated upon but the consequences are really striking.

A comparison of the demographic characteristics of UK and US sites

A very useful way of demonstrating the magnitude of the differences between typical UK and US nuclear sites is to use cluster analysis to provide a multivariate classification of the reactor sites in terms of a set of comparable demographic variables. The resulting classification of sites should provide a concise summary of the different types of reactor sites that have emerged in both countries. The variables used in this analysis are shown in Table 7.4. They are all expressed as population densities, that is persons per square mile. The US data are based on Aldrich *et al.* (1982) while the UK data comes from the 1981 census and was calculated by the author. The form of cluster analysis used here in a non-hierarchical iterative relocation

TABLE 7.4 *Demographic variables used in the multivariate classification of UK and US nuclear reactor sites*

Variable	Description
1	population density within 0-5 miles
2	population density within 5-10 miles
3	population density within 10-20 miles
4	population density within 20-30 miles
5	population density within 0-5 miles in the most populated $22\tfrac{1}{2}$ degree sector
6	population density within 5-10 miles in the most populated $22\tfrac{1}{2}$ degree sector
7	population density within 10-20 miles in the most populated $22\tfrac{1}{2}$ degree sector
8	population density within 20-30 miles in the most populated $22\tfrac{1}{2}$ degree sector
9	site population factor for 5 mile radius multiplied by 1000[a]
10	site population factor for 10 mile radius multiplied by 1000
11	site population factor for 20 mile radius multiplied by 1000
12	site population factor for 30 mile radius multiplied by 1000

Note: a the multiplication by 1000 is to express the site population factor in the same units as the other variables; viz. persons per square mile.

procedure that is widely used to analyse 1981 census data in the UK – see Openshaw (1982, 1983). It is noted that classification is only a quasi-objective form of analysis. A number of critical decisions affect the results; especially, the choice of variables and the selection of an appropriate number of clusters or types. The former was dictated by the nature of the available data whilst the latter is a subjective decision that is based on an analysis of the loss of information as the number of types decrease and on the basic requirement to have as few types as possible in order to obtain a most parsimonious classification. Table 7.5 shows the change in the within cluster sum of squares as the number of clusters or site types reduce from 12 down to 2. This statistic is a measure of the homogeneity of the classification; a value of zero would imply a perfect classification which is never attainable with real data. Most classifications in common use would seem to have a within cluster sum of squares of around 35 per cent.

An approximate rule of thumb that can be applied to Table 7.5 is to look for a 'significant' discontinuity in the within cluster sum of squares function. A major jump occurs between 5 and 6 types suggesting that perhaps 6 is the most parsimonious choice. The reader who is interested further in the taxonomic method that was used should refer to the CCP package program which is described in Openshaw (1982).

Table 7.6 shows the number of sites assigned to each of the 6 types or clusters and Table 7.7 gives the average demographic characteristics of each of them. The cluster numbers have been

TABLE 7.5 *Within cluster sum of squares for alternative site classifications*

Number of clusters	Percentage within cluster sum of squares
2	57.2
3	47.7
4	44.0
5	44.2
6	28.4
7	28.2
8	23.5
9	19.1
10	18.2
11	16.8
12	16.3

TABLE 7.6 *Number of nuclear sites belonging to each type*

Site type	Number of sites	Percentage of US sites	Percentage of UK sites
1	3	2.1	5.5
2	7	1.0	33.3
3	4	3.2	5.5
4	7	5.4	11.1
5	29	27.4	22.2
6	59	60.4	22.2
Total sites		91	18

TABLE 7.7 *Mean demographic characteristics of the six nuclear site types*

| Variable | Nuclear site types | | | | | |
	1	2	3	4	5	6
1	1107	454	397	146	86	34
2	750	491	280	298	124	53
3	650	711	335	417	238	59
4	1523	573	153	827	239	91
5	4919	3177	2138	818	581	282
6	2848	2721	1493	1641	760	378
7	2549	4661	2286	2509	1631	348
8	11327	3023	712	5912	1473	611
9	868	394	346	122	71	27
10	857	429	322	188	88	36
11	768	533	332	263	135	43
12	907	543	296	379	157	53

arranged so that type 1 corresponds to the most heavily populated site and type 6 to the least populated site. Labels can be given to each of the clusters by identifying the common characteristics of their members. The classification appears to be exceptionally interesting and the characteristics and the composition of each cluster are examined in turn.

Site type 1: This cluster is formed by 3 sites; Indian Point (1955), Limerick (1967), and Hartlepool (1968). These sites are located in areas of the highest population densities. In fact these sites have average densities over all four distance bands which are about 8 times greater than the average for all 109 sites. They may well be considered to be almost full urban sites. Within this group, Hartlepool might be regarded as being located in the most densely populated area; indeed, in most of the classifications that were examined it formed a separate cluster by itself because of its highly distinctive density characteristics.

Site type 2: This cluster consists of sites which are located in peri-urban situations; that is in rural areas in close proximity to large cities. The member sites are: Zion (1967) in the USA and Bradwell (1956), Berkeley (1956), Oldbury (1960), Heysham (1970), Druridge Bay (adopted in 1982), and Harwell (1947). These sites are characterised by high population densities in the 10- to 20-mile distance band, about 3 times greater than the average, and densities which are between 3 and 4 times greater in most other distance bands up to 20 miles. The UK sites comprise a mixture of early remote sites (pre-1963) and later (post-1967) relaxed semi-urban sites. It is most interesting that both groups are considered to have similar demographic characteristics and it clearly shows the distorted picture that can be given by the UK siting terminology of 'remote' and 'relaxed' sites. It is also interesting that the US sites included in types 1 and 2 are those which are often quoted as prime examples of sites where either current nuclear operations should be suspended or where no further developments should be allowed because of the very unfavourable demographic characteristics. The inclusion of Druridge Bay in this cluster is very significant because this is currently a greenfield site which has recently been adopted by the CEGB for 2 or more PWRs, yet it would seem that a site with similar demographic characteristics to Druridge Bay would not now be considered suitable for a PWR or any reactor in the US.

Site type 3: This is another group of periurban sites which are located near to small or moderately sized towns or cities. The US sites included here are: Midland (1968), Millstone (1965) and Three Mile Island (1966). The only UK site included is Winfrith (1957). This location has been named for future PWR development and it may also be the site for the first commercial fast breeder reactor in the UK. The characteristic demographic features of this cluster are high population densities within 0 to 5 miles; they are generally 3 times greater than the average for all the sites examined.

Site type 4: The members of this cluster have population densities in the 20- to 30-mile distance band between 2 and 5 times higher than the average. Typically these sites are within this distance of major population concentrations. The US sites are: Baillys (1967), Beaver Valley (1967), Fermi (1955), Haddem

Neck (1962), and Waterford (1970). The UK sites are Hunterston (1956) and Hinkley Point (1957). The latter sites are designated 'remote' sites although they are proximate to Glasgow and Bristol respectively.

Site type 5: This is a collection of 29 sites with slightly less than average population densities in all distance bands. The US sites are: Byron (1971), Catawaba (1972), Cook DC (1967), Dresden (1955), Duane Arno (1968), Forked River (1969), Fort Calhoun (1966), R E Ginna (1965), Marble Hill (1974), W B Mcguire (1969), Oyster Creek (1963), Peach Bottom (1958), T L Perkins (1973), Perry (1972), Pilgrim (1965), Quad Cities (1966), Salem (1966), San Onofre (1963), Sesquoyah (1968), Shoreham (1967), Skagit (1973), Surry (1966), Susquehanna (1968), Turkey Point (1965), W H Zimmer (1969); while the UK sites are: Chapelcross (1955), Windscale (1955), Dungeness (1959), and Sizewell (1960). These UK sites are the most remote of the commercial power station sites.

Site type 6: Finally, cluster 6 is a collection of very remote sites. The population densities are typically 2 to 4 times lower than average for all distances with especially low densities in the 0-to-5 and 10-to-20 distance bands. This is by far the most popular type of US nuclear site. The sites in this group are: Allens Creek (1973), Arkansas (1967), Bellefonte (1970), Big Rock Point (1959), Black Fox (1973), Braidwood (1972), Browns Ferry (1966), Brunswick (1968), Callaway (1973), Calvert Cliffs (1967), Cherokee (1973), Clinton (1973), Commanche Peak (1972), Z Cooper (1967), Crystal River (1967), Davis-Besse (1968), Diablo Canyon (1966), J M Fairley (1969), J A Fitzpatrick (1968), Fort St Vrain (1965), Grand Gulf (1972), Hartsville (1972), E I Hatch (1967), Kewaunee (1967), Lasalle (1970), Lacrosse (1962), Maine Yankee (1967), Monticello (1966), Nine Mile Point (1963), North Anna (1967), Oconee (1966), Palisades (1966), Palo Verde (1973), Peeble Springs (1973), Philipps Bend (1974), Point Beach (1966), Praire Island (1967), Rancho Seco (1967), River Bend (1972), H B Robinson (1966), Saint Lucie (1976), Seabrook (1972), Shearon Harris (1971), South Texas (1973), V C Summer (1971), Trojan (1968), Vermont Yankee (1966), A W Vogtle (1971), Watts Bar (1970), WPSS1+4 (1972), WPSS3+5 (1973), WPSS2 (1971), Wolf Creek (1973), Yankee Rowe (1956), and Yellow Creek (1974). The following UK sites are included in this cluster: Dounreay (1954), Trawsfynydd (1958), Wylfa (1961), and Torness (1979).

Interpretation: The results are somewhat surprising since the a priori expectation was that the UK and US sites would not have similar demographic characteristics because of underlying

257

differences in population density and in siting practices. However, this is not the case, and none of the alternative classifications that were examined displayed this characteristic. Instead the differences between the two countries are manifest in the different proportions of sites that occur in each of the six types as shown in Table 7.6. Most of the US sites (60 per cent) are located in areas with very low population densities and most of the remainder (a further 27 per cent) in areas of average or lower than average densities. The equivalent percentages of UK sites are 22 and 22 respectively; or about one half of the frequency of the US sites. This is partly due to the fact that in the 1960s virtually all US applications for metropolitan sites were rejected or subsequently abandoned whilst in the UK the only two applications were accepted. However, most of the differences result from the restricted range of the UK's remote siting criteria; distances out to 10 miles only were considered. This has allowed 'remote' sites to be found in close proximity to major population concentrations; indeed, as the analyses in Chapter 3 show a large number of such sites exist although, perhaps ironically, were not found at the time. Finally, it is noted that it is difficult to identify any trends through time. The dates in parentheses after the various site names are the order dates for the first reactor located there – see Coffin (1984). This is because most of the US sites were selected in a single 10-year period whilst the UK sites span a longer time interval and have far more erratic demographic characteristics. For example, the Druridge Bay site would satisfy the Marley and Fry (1955) remote siting criteria even though it will be tested against the current relaxed siting criteria. However, as the cluster analysis also shows such designations have little relevance as a description of a site's demographic characteristics.

Conclusions

This chapter has considered some of the technical problems associated with the production of demographic data for nuclear reactor sites and has described a set of basic statistics which may be of general interest. It is noted that too strong an emphasis should not be placed on the absolute accuracy of census-based population counts, or indeed on the accuracy of estimates obtained by any other means. What this suggests is that sites which are known to be marginal, or within some modest percentage of limiting population values (say 10 per cent) may well exceed the threshold values if more accurate demographic data were to be obtained. Additionally, it is no use obtaining accurate counts of people for fixed sectors if an optimal rotation will yield different conclusions – for example, to identify the maximum population sector. Moreover, the type of population, its local distribution in terms of its residential characteristics, and local infrastructure features must also be taken into account at some stage when deciding on the acceptability of a nuclear site (private correspondence with Mr G. Johnston, CEGB Station Planning Engineer). What this means is that sites which are acceptable in terms of their total population numbers may on closer examination be found to be unacceptable for reasons not readily apparent from census data. However, this is a poor excuse, if it were to be used for the purpose, for failing to apply available computer search methods based on demographic data bases. It can also be argued that such considerations will in any case mainly affect sites which are so marginal with respect to the demographic criteria that they should really have never been considered in the first place. Furthermore, these and other qualitative judgemental aspects of the site selection process properly belong to a later stage of analysis.

The marginal nature, by US standards, of many of the UK's nuclear sites should be a cause for some concern. Are we really so hard up for good sites that nothing 'better' can be found in areas of lower population densities? Or is it that we have never either had a proper look or given much emphasis to siting aspects of nuclear safety. Until fairly recently there was a good excuse for this neglect. UK reactor designs were sufficiently different from those used in other countries to justify the claim, which has never been substantiated, that British reactors are a breed apart in terms of their theoretical and inherent safety characteristics. However, as the UK adopts US reactor technology it is fair to enquire whether we should not also adopt US site evaluation methods and even the US style of siting criteria. Certainly vast improvements have been made in the application of computer-based search methods to the identification and evaluation of nuclear sites; see for example, Dobson (1979, 1983), and Hobbs and Voelker (1977). The nuclear plant planning engineer who continues to neglect these new methods of analysis and computer assisted searches can be fairly criticised for an undue concentration on manual methods and informal judgements. This criticism would seem to be applicable to past and present UK siting practices. It has happened because of the long pre-eminence of engineering factors, with the result that the full impact of the various demographic criteria have not been identified until recently (Openshaw, 1982b) and the full range of potential sites have never been input or included in the evaluation process. If

259

nuclear power is to remain publicly acceptable as an energy option of anything other than last resort, then it is important that a more realistic site investigation, identification, and evaluation process based on widely available computer and information processing techniques be used. An important first step would be a greater emphasis on the analysis of demographic data to be backed up by the provision of other necessary data bases for computer analysis.

Appendix 4

TABLE 1 *Total populations for selected distances around UK and US reactor sites*

| Site | Country | Total population ('000s) within | | | | | |
		5	10	20	30	50	100 miles
Allens Creek	US	2	4	28	61	1437	1130
Arkansas	US	4	19	24	25	75	989
Baillys	US	21	66	503	1608	4554	3416
Beaver Valley	US	12	133	322	1236	2025	4948
Bellefonte	US	1	20	28	64	738	2049
Berkeley	UK	21	105	859	1021	2181	15880
Big Rock Point	US	4	3	25	14	80	259
Black Fox	US	2	2	138	367	180	895
Bradwell	UK	11	118	704	926	6574	9519
Braidwood	US	9	12	74	263	3518	6078
Browns Ferry	US	0	28	82	153	356	1790
Brunswick	US	2	5	58	40	65	942
Byron	US	6	13	235	199	427	10343
Callaway	US	0	2	30	136	120	2898
Calvert Cliffs	US	2	12	51	80	2292	4735
Catawaba	US	3	55	406	241	537	2733
Chapelcross	UK	15	13	163	102	266	8152
Cherokee	US	3	26	106	345	814	2238
Clinton	US	1	10	33	263	397	1602
Commanche Peak	US	1	4	6	51	713	2214
Cook DC	US	7	36	108	354	588	9848
Cooper	US	1	5	17	34	110	1649
Crystal River	US	1	7	10	12	155	2097
Davis-Besse	US	2	12	83	596	1065	8246
Diablo Canyon	US	0	7	65	50	85	306
Dounreay	UK	0	10	5	12	25	282
Dresden	US	5	27	187	406	5815	3675
Duane Arnold	US	3	81	39	58	271	1366
Dungeness	UK	6	12	242	345	2065	13053
Druridge Bay	UK	43	97	674	724	839	3557
Fairley	US	1	6	66	42	206	1130

TABLE 1 *cont.*

Site	Country	Total population ('000s) within					
		5	10	20	30	50	100 miles
Fermi	US	9	61	363	1969	2824	4571
Fitzpatrick	US	2	35	47	113	648	1861
Forked River	US	5	30	137	276	2839	20616
Fort Calhoun	US	7	5	294	285	115	801
Fort St Vrain	US	0	8	134	295	965	353
Grand Gulf	US	1	6	17	62	201	1154
H B Robinson	US	7	17	47	117	387	2309
Haddem Neck	US	8	49	445	1261	1533	19367
Hartlepool	UK	139	295	520	1014	1020	9424
Hartsville	US	3	8	57	72	743	1083
Harwell	UK	36	98	688	1010	6343	19815
Hatch	US	1	4	35	43	165	966
Heysham	UK	60	86	482	659	4659	10414
Hinkley Point	UK	5	73	459	1074	2201	6008
Hunterston	UK	8	79	391	1176	929	1884
Indian Point	US	59	145	689	3213	12375	7162
Kewaunee	US	1	7	75	155	331	1979
La Crosse	US	1	5	83	53	175	1295
La Salle	US	0	12	84	117	703	9212
Limerick	US	62	89	629	2948	3111	16611
Maine Yankee	US	0	1	33	98	226	424
Marble Hill	US	6	10	283	595	336	3322
Mcguire	US	5	32	475	303	567	2615
Midland	US	42	20	272	133	547	4358
Millstone	US	45	66	157	160	2060	14702
Monticello	US	5	8	42	243	1709	824
Nine Mile Point	US	2	35	47	113	648	1861
North Anna	US	0	6	27	91	733	4311
Oconee	US	3	41	64	256	361	1814
Oldbury	UK	34	138	959	1206	1794	14419
Oyster Creek	US	5	30	137	276	2839	20616
Palisades	US	5	24	86	91	794	9966
Palo Verde	US	0	1	7	10	613	424
Peach Bottom	US	3	22	231	568	3312	10084
Pebble Springs	US	0	0	0	3	75	353
Perkins	US	6	25	191	394	864	2261
Perry	US	17	54	167	464	1879	3180
Philipps Bend	US	6	13	120	153	392	1837
Pilgrim	US	9	20	124	639	3513	2591
Point Beach	US	2	18	59	138	351	2120
Prairie Islands	US	4	15	48	179	1799	1083
Quad Cities	US	1	15	294	120	236	2002
R E Ginna	US	6	29	575	224	336	2686
Rancho Seco	US	1	6	125	772	467	4948

TABLE 1 *cont.*

Site	Country	**Total population ('000s) within**					
		5	10	20	30		100 miles
River Bend	US	3	17	81	276	216	2167
Saint Lucie	US	5	37	32	45	206	1366
Salem	US	3	24	314	546	3910	9660
San Onofre	US	1	24	172	210	3176	7398
Seabrook	US	9	20	83	100	1367	3039
Sequoyah	US	8	27	285	111	256	1932
Shearon Harris	US	1	16	158	322	547	2285
Shoreham	US	10	34	327	1330	3513	16823
Sizewell	UK	10	19	148	375	824	11757
Skagit	US	3	12	32	103	216	1743
South Texas	US	0	2	23	17	130	2214
Surry St	US	2	59	174	304	1065	942
Susquehanna	US	14	30	311	279	864	8906
Three Mile Island	US	25	110	470	389	844	11922
Torness	UK	4	5	43	348	1045	5207
Trawsfynydd	UK	7	10	62	175	924	12110
Trojan	US	8	46	47	81	955	1130
Turkey Point	US	0	38	168	686	764	612
Vermont Yankee	US	8	18	93	106	1090	8552
Virgil C Summer	US	0	10	44	304	336	2591
Vogtle	US	0	1	24	254	175	1366
Waterford	US	14	28	265	769	457	942
Watts Bar	US	1	7	57	106	507	1437
Windscale	UK	13	41	81	136	678	12016
Winfrith	UK	11	65	365	282	1462	8717
Wolf Creek	US	2	0	8	50	105	2285
WPPSS1+4	US	0	1	65	34	80	329
WPPSS2	US	0	1	57	42	80	329
WPPSS3+5	US	2	5	43	83	246	2026
Wylfa	UK	4	21	51	78	165	7563
Yankee Rowe	US	0	20	79	202	1281	7327
Yellow Creek	US	1	7	39	54	246	1555
Zimmer	US	4	20	191	977	633	3675
Zion	US	42	164	327	760	5679	4618

TABLE 2 *Populations in the most populated $22^1{}_2$ degree sectors for selected distances around UK and US reactor sites*

Site	Country	Total population ('000s) within			
		5	10	20	30 miles
Allens Creek	US	16	42	123	240
Arkansas	US	28	159	105	109
Baillys	US	88	388	3876	14599
Beaver Valley	US	84	496	946	9737
Bellefonte	US	15	99	83	125
Berkeley	UK	135	280	6999	3662
Big Rock Point	US	56	11	151	45
Black Fox	US	20	19	1082	3506
Bradwell	UK	80	693	3492	4140
Braidwood	US	48	66	385	2297
Browns Ferry	US	14	191	473	1147
Brunswick	US	35	26	763	400
Byron	US	27	40	2065	558
Callaway	US	10	13	152	875
Calvert Cliffs	US	23	56	207	269
Catawaba	US	20	380	2563	953
Chapelcross	UK	132	66	1319	910
Cherokee	US	21	231	422	1268
Clinton	US	8	67	78	1572
Commanche Peak	US	24	20	27	288
Cook DC	US	26	248	447	3032
Cooper	US	4	25	59	131
Crystal River	US	18	38	48	82
Davis-Besse	US	26	74	393	3703
Diablo Canyon	US	0	41	534	464
Dounreay	UK	5	144	39	147
Dresden	US	26	84	1907	1717
Duane Arnold	US	21	586	96	135
Dungeness	UK	51	123	1327	1466
Druridge Bay	UK	372	796	5955	7394
Fairley	US	12	31	584	72
Fermi	US	46	321	2485	10299
Fitzpatrick	US	36	414	292	941
Forked River	US	36	202	798	1617
Fort Calhoun	US	76	56	3027	2503
Fort St Vrain	US	10	28	541	1516
Grand Gulf	US	16	39	57	472
H B Robinson	US	41	123	187	412
Haddem Neck	US	62	207	1626	4288
Hartlepool	UK	629	1245	2937	11103
Hartsville	US	35	18	258	251
Harwell	UK	233	556	3199	3613

TABLE 2 *cont.*

| Site | Country | Total population ('000s) within | | | |
		5	10	20	30 miles
Hatch	US	16	26	128	96
Heysham	UK	572	487	3563	4405
Hinkley Point	UK	32	557	2774	4782
Hunterston	UK	67	627	1644	13341
Indian Point	US	197	451	2227	22961
Kewaunee	US	17	46	767	2030
La Crosse	US	11	16	840	252
La Salle	US	9	45	361	530
Limerick	US	332	315	2042	19315
Maine Yankee	US	0	11	206	1073
Marble Hill	US	50	39	2184	5408
Mcguire	US	30	100	2918	680
Midland	US	157	65	2093	477
Millstone	US	293	322	815	394
Monticello	US	35	44	92	975
Nine Mile Point	US	36	414	292	941
North Anna	US	14	23	53	462
Oconee	US	16	193	261	1446
Oldbury	UK	192	645	5973	4773
Oyster Creek	US	36	202	798	1617
Palisades	US	32	108	889	346
Palo Verde	US	5	12	71	138
Peach Bottom	US	22	60	1218	1716
Pebble Springs	US	6	4	4	13
Perkins	US	36	74	636	1272
Perry	US	63	367	847	6027
Philipps Bend	US	20	67	863	875
Pilgrim	US	69	144	389	2785
Point Beach	US	27	206	581	982
Prairie Islands	US	22	140	206	1361
Quad Cities	US	8	56	1826	602
R E Ginna	US	54	121	5544	1100
Rancho Seco	US	27	23	540	4849
River Bend	US	23	70	414	2628
Saint Lucie	US	74	318	208	476
Salem	US	49	141	1898	2463
San Onofre	US	22	209	1000	1967
Seabrook	US	42	110	517	712
Sequoyah	US	23	87	1790	431
Shearon Harris	US	14	57	679	1737
Shoreham	US	63	192	1498	5057
Sizewell	UK	76	76	1007	2360
Skagit	US	22	123	195	789
South Texas	US	0	14	250	83

TABLE 2 *cont.*

Site	Country	Total population ('000s) within			
		5	10	20	30 miles
Surry St	US	19	412	1244	2389
Susquehanna	US	102	132	2413	1366
Three Mile Island	US	169	546	1529	1819
Torness	UK	50	64	263	3785
Trawsfynydd	UK	67	87	274	1079
Trojan	US	28	506	166	915
Turkey Point	US	0	303	1986	6471
Vermont Yankee	US	39	125	340	550
Virgil C Summer	US	1	23	194	3073
Vogtle	US	0	17	72	1557
Waterford	US	69	106	3203	7960
Watts Bar	US	15	23	233	256
Windscale	UK	128	517	754	1333
Winfrith	UK	51	473	4181	1787
Wolf Creek	US	33	5	15	353
WPPSS1+4	US	0	22	548	248
WPPSS2	US	0	22	507	310
WPPSS3+5	US	35	45	509	354
Wylfa	UK	26	179	439	599
Yankee Rowe	US	7	166	269	1053
Yellow Creek	US	10	23	247	160
Zimmer	US	25	42	894	8374
Zion	US	160	1029	1569	5253

TABLE 3 *Site population factors for UK and US reactor sites*

Site	Country	Site population factors			
		5	10	20	30 miles
Allens Creek	US	31	26	27	29
Arkansas	US	34	60	48	41
Baillys	US	171	214	333	462
Beaver Valley	US	90	250	288	386
Bellefonte	US	60	72	58	54
Berkeley	UK	242	331	544	574
Big Rock Point	US	32	25	27	23
Black Fox	US	17	14	55	93
Bradwell	UK	149	262	442	475
Braidwood	US	135	109	96	113
Browns Ferry	US	7	44	64	70
Brunswick	US	20	22	33	31

265

TABLE 3 *cont.*

Site	Country	Site population factors			
		5	10	20	30 miles
Byron	US	71	67	118	120
Callaway	US	9	10	19	32
Calvert Cliffs	US	19	30	40	42
Catawaba	US	28	97	201	193
Chapelcross	UK	224	154	163	139
Cherokee	US	32	60	74	104
Clinton	US	19	31	33	62
Commanche Peak	US	8	12	10	15
Cook DC	US	84	113	119	140
Cooper	US	10	14	16	17
Crystal River	US	16	22	18	16
Davis-Besse	US	32	40	56	121
Diablo Canyon	US	0	9	35	34
Dounreay	UK	18	27	19	16
Dresden	US	44	67	117	143
Duane Arnold	US	39	129	128	109
Dungeness	UK	54	55	126	147
Druridge Bay	UK	368	391	477	481
Fairley	US	11	17	33	32
Fermi	US	158	191	245	440
Fitzpatrick	US	19	68	62	63
Forked River	US	80	94	112	125
Fort Calhoun	US	73	55	125	140
Fort St Vrain	US	7	20	63	86
Grand Gulf	US	12	19	19	23
H B Robinson	US	44	60	56	59
Haddem Neck	US	122	149	244	365
Hartlepool	UK	1095	1249	976	899
Hartsville	US	21	26	37	39
Harwell	UK	376	398	519	546
Hatch	US	11	14	22	23
Heysham	UK	646	544	530	503
Hinkley Point	UK	47	155	274	362
Hunterston	UK	105	209	273	384
Indian Point	US	813	740	735	986
Kewaunee	US	9	17	36	47
La Crosse	US	17	19	39	38
La Salle	US	13	29	55	59
Limerick	US	695	581	592	837
Maine Yankee	US	0	1	14	23
Marble Hill	US	52	48	123	180
Mcguire	US	68	89	222	218
Midland	US	516	362	328	278
Millstone	US	445	397	319	274

TABLE 3 *cont.*

Site	Country	Site population factors			
		5	10	20	30 miles
Monticello	US	36	36	40	63
Nine Mile Point	US	19	68	62	63
North Anna	US	7	16	20	28
Oconee	US	20	71	71	87
Oldbury	UK	265	372	640	665
Oyster Creek	US	80	94	112	125
Palisades	US	54	74	78	74
Palo Verde	US	5	5	6	6
Peach Bottom	US	21	46	103	154
Pebble Springs	US	3	2	1	1
Perkins	US	56	73	119	147
Perry	US	181	196	187	207
Philipps Bend	US	105	84	97	97
Pilgrim	US	115	109	115	172
Point Beach	US	27	42	50	56
Prairie Islands	US	52	60	58	68
Quad Cities	US	9	28	115	108
R E Ginna	US	47	72	235	217
Rancho Seco	US	11	16	49	142
River Bend	US	30	43	55	81
Saint Lucie	US	54	83	69	61
Salem	US	20	44	125	170
San Onofre	US	6	48	88	96
Seabrook	US	67	70	75	74
Sequoyah	US	74	92	164	145
Shearon Harris	US	19	32	71	100
Shoreham	US	34	43	42	47
Sizewell	UK	159	127	130	158
Skagit	US	164	158	221	353
South Texas	US	0	3	11	11
Surry St	US	11	101	126	140
Susquehanna	US	88	107	179	179
Three Mile Island	US	229	312	399	371
Torness	UK	40	35	39	76
Trawsfynydd	UK	69	58	58	70
Trojan	US	60	107	88	80
Turkey Point	US	0	53	98	167
Vermont Yankee	US	95	94	95	88
Virgil C Summer	US	0	16	26	59
Vogtle	US	0	3	10	39
Waterford	US	163	149	186	255
Watts Bar	US	15	22	34	41
Windscale	UK	141	149	123	116
Winfrith	UK	192	218	282	261

TABLE 3 *cont.*

Site	Country	Site population factors			
		5	10	20	30 miles
Wolf Creek	US	16	12	11	15
WPPSS1+4	US	0	3	25	24
WPPSS2	US	11	18	27	31
WPPSS3+5	US	0	2	22	23
Wylfa	UK	79	81	71	66
Yankee Rowe	US	12	35	51	67
Yellow Creek	US	6	14	22	25
Zimmer	US	27	46	93	206
Zion	US	713	706	581	556

8 Optimal siting from a public perspective

The previous chapters largely support the conclusion that nuclear power station siting practices are very much a reflection of an engineer's approach to the problem. Demographic criteria and amenity factors exist as constraints which are imposed on the engineer's solution space and their severity and rigorousness depend on the cultural traditions of the countries involved. In both the US and the UK the demographic siting criteria cannot be too stringent for all kinds of reasons, yet siting is the only reactor design independent safety measure that does not depend on institutions for successful operation. However it is not clear to what extent the current demographic criteria are useful as additional safeguards to the public. The US concept of 'defence in depth' is a good idea but perhaps the existing sites are far too close to major population concentrations to offer any real benefits. It looks as if the use of remote sites is becoming increasingly unnecessary because engineering safety measures are probably more cost effective and continued improvements render any other safeguards irrelevant except for worst-case accidents.

In the UK the choice of site is generally regarded by the safety agencies as offering a tertiary level of protection. Dunster and Clarke (1980) argue that the differences in levels of consequence between the best and the worst sites are considerably smaller than the protection that is provided by the engineered safety systems of the plant. Whilst true, it is also a meaningless statement since no sane person would recommend remote siting as a substitute for engineered safety measures. The argument is merely whether there is a strong safety case for remote siting. Dunster and Clarke (1980) also state 'A reactor siting policy which favoured remote sites would not increase the safety of reactors – it might marginally decrease that safety' (p. 53). The justification is that remote sites may suffer from a greater

269

frequency of unplanned disconnections from the grid because of the longer transmission lines involved. These unscheduled events might stress the safety systems, and if other failures occur simultaneously an accident might result. One hopes that this is in fact a very poor argument. Nevertheless, whilst the benefits of real remote siting may be less than commonly thought it is strange that the public safety agencies should be arguing against it – a point of view that might be more readily associated with the power utilities themselves.

The main redeeming feature of the Dunster and Clarke (1980) paper is their belief that the means are now available to provide a quantitative basis for reviewing the radiological aspects of potential reactor sites in a much more rigorous manner than previously. Indeed they suggest that consequence calculations could be used to produce a preference ranking of sites, an approach which might well result in the identification of minimum casualty locations. Openshaw (1982b) has already adopted this approach and provides a relative assessment of the number of potential coastal sites in the UK with lower casualties than each of the existing nuclear power sites. For this purpose two severe accident scenarios were examined; one for the PWR based on PWR2 from NRC (1975), the other for an AGR from PERG (1980). Table 8.1 shows the number of lower casualty locations with less than 50 per cent of the total casualties (early deaths and latent cancers combined) for each power station site. While many of these lower casualty locations may be unsatisfactory locations for nuclear power stations, it would really extend credulity if they were all unsuitable.

So one approach to ensuring public safety would be to seek optimally remote sites as measured by reference to the numbers of casualties likely to occur should a particular reactor accident occur. However, the arguments in favour of this form of optimal siting policy are weak; because of uncertainties in accident source terms, large amounts of variability due to assumptions about weather conditions, and the very small probabilities of occurrence. It is difficult to separate consequences from probabilities of occurrence. The most likely accident scenario will have no public health consequences, whilst those that do are predicted to be exceptionally rare events. So this argument could be used to justify urban siting on the grounds that, all things being equal, these are optimal minimum casualty locations.

Health consequence predictions

Another indication of the impending demise of siting as a safety

TABLE 8.1 *Percentage of UK coastal grid-squares with 50 per cent lower casualties than existing nuclear power stations*

Site	PWR accident	AGR accident
Berkeley	53	47
Bradwell	70	55
Calder Hall	43	36
Chapelcross	30	28
Dounreay	9	9
Dungeness	55	41
Hartlepool	57	50
Heysham	61	50
Hinkley Point	48	41
Hunterston	68	58
Oldbury	47	44
Sizewell	54	42
Torness	42	36
Trawsfynydd	38	29
Winfrith	54	46
Wylfa	32	28

Source: Openshaw (1982b).

measure can be seen in the latest casualty estimates that have been made in both the US and the UK. It would seem that now only the most incredibly severe accidents will produce large health effects.

For the purposes of decision-making such as siting and emergency response planning the NRC have defined five release scenarios to represent the range of potential accidents (NRC, 1982). The spectrum includes: SST1 (siting source term 1) which assumes severe core damage with the loss of all safety features and the direct breach of the containment, in other words a class 9 accident or an accident beyond the design basis; SST2 which assumes severe core damage but the fission product release is mitigated by the safety systems; SST3 severe core damage with containment failure by basemat meltthrough but all other release mitigation systems work; SST4 which assumes limited to moderate core damage with containment systems operating to some extent; and SST5 with limited core damage and all safety measures working as designed. Table 8.2 gives an indication of the mean health consequences for a 1120MW(e) PWR reactor located at Indian Point with New York meteorology.

It is interesting that the offsite consequences for the SST4 and 5 accidents are zero even for Indian Point which is the most

TABLE 8.2 *Mean health effects for various siting source terms at Indian Point*

Accident scenario	Early fatalities	Early injuries	Latent cancer fatalities
SST1	831	3640	8110
SST2	0	18	587
SST3	0	0	2
SST4	0	0	0
SST5	0	0	0

Source: Strip (1982), p. 4.

densely populated of all US sites and quite typical of worst case UK sites. The mean consequences for SST1 exceed those for SST2 by between 1 and 4 orders of magnitude and those for SST3 to 5 by between 4 and 7 orders of magnitude (Strip, 1982; Aldrich *et al.*, 1982). These results are obtained with source terms which are now widely believed to be too large and that certain inventories need to be reduced by an order of magnitude to be realistic (NRC, 1982). These reductions reflect a better understanding of what is likely to happen to radionuclides from a core inventory during an accident. The effect would be to reduce early fatalities by about two orders of magnitude and other consequences by one order of magnitude. So it would seem that increasingly only the worst imaginable highly improbable accidents would have any potential for significant health effects.

There is also a slight risk in only considering mean or average consequences. The 99 percentile limits for the SST1 casualties given in Table 8.2 are large; for example, the mean of 831 early fatalities has a 99 percentile limit of 8,200; the mean number of early injuries has a limit of 33,000 and latent cancers of 24,000. Moreover, these limits could be exceeded – for example, if it were to rain over a heavily populated area or if the SST1 source term turned out to be optimistic. Nevertheless, the role of siting as a safety measure is perhaps being steadily reduced by a better understanding of the mechanisms of dispersal under accident conditions.

Geographical consequence predictions

Apart from health effects the geographical consequences of an accident might also be reduced. Table 8.3 gives estimates of the distances of various events associated with reactor accidents. For the SST1 the geographical range over which protective actions are needed is large, extending over 50 miles; for SST2 to 30 miles, and 3 miles for SST3.

TABLE 8.3 *Ranges of effects due to reactor accidents*

Accident scenario	Mean distances (miles)		
	Fatal injuries	injuries	interdiction
SST1	3.9 (12)	11 (35)	19 (55)
0.5* SST1	2.5 (10)	7 (20)	14 (45)
0.1* SST1	.9 (2)	3 (10)	6 (18)

Note: Values in parentheses are 99 percentile limits.
Source: Aldrich *et al.* (1982).

This need to evacuate quite large areas, including several major urban areas, might be regarded as a more profound consequence than the direct health effects. There could well be problems of decontamination and the possibility of long term evacuations (in the order of 30 to 50 years) and restrictions on land use. Little is known about the effects of the deposition of radioactive material on residential and industrial areas, and if evacuations were not performed then the health effects could be considerably greater – especially latent cancers.

Financial consequence predictions

Another important aspect of reactor accidents is their likely financial implications. This is relevant to the safety debate because the financial implications are being used as a basis for evaluating the cost-effectiveness of both proposed design changes and backfit proposals. In the US the expiry of the limited liability Price-Anderson Act has given utilities a major financial incentive to co-operate with the NRC in seeking the safest possible sites (Starr and Whipple, 1982). In addition the safety principles of either 'as low as reasonable practicable' (ALARP) or 'as low as reasonably possible' (ALARP) used in the US and the UK tend to define reasonableness in both technical and financial terms. So it would seem that value-impact analyses are likely to play an important role in decision-making processes especially those related to 'big' accidents.

Strip (1982) provides some estimates of the likely magnitude of the financial consequences of the five site source terms for health

273

effect costs, property, and on-site costs of replacement power and clean-up. Table 8.4 shows the costs of various accidents at Indian Point. These financial estimates assume a value of 1 million dollars for each early fatality (the value currently used to evaluate additional safeguards) and 0.1 million dollars for early injuries and latent cancers; both exclude the costs of medical care and possible litigation.

TABLE 8.4 *Financial costs of reactor accidents at Indian Point*

Accident scenario	Early fatalities	Injuries	Mean cost (in million dollars) of: Latent cancers	Property damage	On-site costs	Total
SST1	647	284	632	9200	6430	17200
SST2	.1	1.4	45.8	114	6430	6590
SST3	0	0	.2	15	6430	6450

Note: No estimates available for SST4 and SST5. The plant may be repairable but no cost estimates are available; there are no off-site costs.
Source: Strip (1982), pp. 7, 10.

It is clear that based on the costs in Table 8.4 it would only require one 'big' SST1 type of accident to bankrupt almost any utility, or for that matter the CEGB were they to face the full costs of an accident; remember the CEGB still have a limited liability of 6.3 million dollars in any one accident. It should also be noted that these estimates are by no means the worst that could occur; only the average predictions. The values are also conditional on the accident occurring. Now if they were to be corrected for both a proper discounting of the monetary values and for the accident frequency rate, then the results in Table 8.5 would be obtained. The former greatly increases the costs whilst the latter, if used to compute mean life time risks, has the effect of making the cost of the accidents almost zero. For example, if accident frequencies of 10^{-6} per year were to be assumed (in the UK values of 10^{-9} and smaller are being used for severe accidents) then the total mean risk would be 281,000 dollars per year. Perhaps the leading question is whether the utilities and the insurance companies really believe or put much faith in, calculations of this sort. It would seem that the real value will probably be either zero or some infinitely large number!

The CEGB would respond to Table 8.4 by emphasising that Table 8.5 is more realistic especially if the frequencies of occurrence are set at some vanishingly small number. They might also point out the value of life in the UK is less in monetary

TABLE 8.5 *Discounted present values of costs of a reactor accident at Indian Point*

Accident scenario	Mean cost (in million dollars) of:					
	Early fatalities	Injuries	Latent cancers	Property damage	On-site costs	Total
SST1	$f_1 \times 11700$	$f_1 \times 5120$	$f_1 \times 11400$	$f_1 \times 16600$	$f_1 \times 86600$	$f_1 \times 281000$
SST2	$f_2 \times$ 1	$f_2 \times$ 25	$f_2 \times$ 826	$f_2 \times$ 2050	$f_2 \times 86300$	$f_2 \times$ 89200
SST3	$f_3 \times$ 0	$f_3 \times$ 0	$f_3 \times$ 2	$f_3 \times$ 274	$f_3 \times 86300$	$f_3 \times$ 86600

Notes: f_1, f_2, f_3 are the frequencies of the accidents occurring.
Source: Strip (1982), pp. 10, 11.

terms than in the US but that the real lesson from these financial impact predictions is the fundamental need to ensure that 'big' accidents never happen by a combination of good management and reactor engineering. It is all a matter of faith and confidence in the power station operators and the absence of unexpected events. If the predictions in Table 8.5 are thought reasonable then clearly the additional costs of remote siting, or any costs related to siting as a safety measure, would not be judged worthwhile. On the other hand if a pessimistic view were to be adopted, the potential costs that might have to be met by the state are so large that virtually any measures to reduce the location dependent component (the health and property costs) might well seem justified. The gamble only looks a good one whilst no large accidents have occurred but there is real uncertainty about the prospects both within individual countries and for the world as a whole of this state continuing indefinitely. As more reactors are built so the odds in favour of a major accident somewhere continue to look more and more certain: it is, from a probabilistic point of view only a matter of time; it may be one year, it may be a million and the probability assessments provide no real guide about the likely date of the first occurrence other than that all dates are equiprobable.

Kemeny report

The situation after a very low probability accident has occurred should of course be very different, although public relations measures may well be used to launder any residual or long-term problems. Three Mile Island is a good example of an accident which might well have been attributed a zero probability of occurring had anyone thought about it. Yet because there was no noticeable health effect the occurrence of what by any probability assessments was an impossible event has been effectively ignored. In the UK the accident has been presented as due largely to operator error; CEGB operators are all trained graduates which is supposed to indicate that they do not make errors. Additionally, the fact that a partial meltdown had caused no directly attributable health consequences has been used to indicate how safe nuclear power really is; the financial cost implications which according to Table 8.4 might come to about 8 billion dollars, are ignored. As a result there has been no need to implement any of the subsequent Kemeny Report recommendations in the UK (see Kemeny, 1980). The Health and Safety Executive, that independent government department responsible for nuclear safety standards, noted:

> The comments in the report are not relevant to the situation in Britain and Europe more generally and do not cast doubts on the validity of the choice of existing British sites (HSE, 1981, p. 21).

No reasons were given. Kemeny merely argued that Three Mile Island demonstrated the need for new sites to be located to the maximum possible extent in areas remote from concentrations of population. It would seem that the strength of this particular recommendation has now been watered down in the US also.

Kemeny also criticised the inadequacies of the Three Mile Island emergency plan which only dealt with a 3-mile area around the reactor when long distance evacuations were being recommended. Once again this is not viewed as relevant to the UK where indeed emergency plans only allow for evacuation out to 2 miles and there is virtually no possibility whatsoever of speedy 10- or 20-mile evacuations. Openshaw (1982b) suggests that perhaps the HSE have been misled about the possible geographical range that emergency plans for evacuation should seek to cover; alternatively, maybe they are merely being realistic because for many UK sites larger evacuation distances would be infeasible due to the numbers of people that would be involved. They may even believe the nuclear industry's estimates, in which case it is hardly worth planning for events which are not likely to occur until well after the next 10 ice ages – for which no preparations have been made either.

Towards a new perspective on the role of siting

An alternative approach is to start with the basic argument that nuclear power station siting is not acceptable as a purely

276

engineering process. The locational decisions have to satisfy not merely technical criteria, but also less well-defined environmental, socioeconomic, and safety constraints. Such a perspective is already implicit in the CEGB's notion of a balanced judgement as well as in the US, even if the weighting factors attributable to the various components are not properly defined, if at all. All too often in the UK the environmental aspects only become important as part of the fine tuning of the design work after the important decisions have been made, whilst the socioeconomic aspects reflect an attempt to mitigate any adverse impacts on the local communities. They may be regarded largely as damage limiting measures to be applied retrospectively to meet statutory requirements that have previously not been satisfied. Such a 'balanced' approach will frankly only be viable if either all the decisions are made in secret so there is no probing public examination (it looks good to talk about balanced decisions made 'in the round' after considering all relevant factors etc.) or if there were a rigorous scientific basis for the comparative evaluations of alternative sites, which are surely needed for a good and balanced judgement. It would seem that this scientific approach to site selection is possible and it has been used in the EEC and the US (see Andreas et al., 1979). It would seem that the essential minimum qualification for a scientific approach is the use of some kind of multiattribute evaluation procedure applied to a large sample of potential sites selected from a wide range of alternative locations.

Of course the expense of performing a detailed and really rigorous scientific evaluation of alternative sites can easily be justified because of the size and the national importance of the subsequent investments. The problem is what criteria do you use for the site assessments and what standards do you set for acceptable sites? Gammon (1979) notes that the final choice reflects both the selection of sites to be evaluated and the balancing of their benefits against the costs and detriments that society will be prepared to accept. The problem here is who is qualified to make these decisions? Can they reasonably be entrusted to the developer, as indeed they are in the UK? How can the CEGB be expected to judge what society will and will not be prepared to accept? The situation in the US is not very different. It is like asking the nuclear industry to decide, in the national interest, whether nuclear power is necessary; they are not going to say no!

As a result of a non-independent site evaluation process, some might say biased, it would seem that the best sites from a public safety point of view have not been selected; nor are they going to be because site related safety criteria are not becoming popular and they rate low on any pragmatic scale of engineering convenience. The argument is therefore that what masquerades as a balanced judgement is in fact a grossly unbalanced one.

These arguments are however largely irrelevant and impotent. The safety case is too weak to gain the necessary political support for the sorts of changes that would be necessary to force remote siting on an unwilling utility. Instead it can be suggested that the most important aspect of the siting debate has not yet surfaced. Quite simply, to survive nuclear power must be accepted not just by governments or the power utilities or the safety agencies or the nuclear industry, but by the public at large. Not only must it be acceptable when the decisions are made to proceed, but it must also remain acceptable for the indefinite future which includes both the current accidentless honeymoon and the subsequent period when accidents have occurred and more are expected. A similar argument but with a different solution has been made by Burwell et al. (1979). They observe that '. . . nuclear matters have acquired such sensitivity, a failure anywhere in the system is likely to affect confidence in the entire system' (p. 1043). The siting decision is important because it can be seen as influencing the acceptability of nuclear power because it affects the number of people who are at risk, as well as in the event of an accident the number of likely casualties. However, it is perhaps the former which is the most important factor in influencing public acceptance of nuclear power at a time when there are alternative energy options available to society. It is the number of people who perceive themselves as being threatened that is the crucial factor in seeking optimal sites, and of course people's perceptions seldom, if ever, have any close resemblance to reality – when there is a reality to compare them with. Remote siting is important therefore as a means of buying the long-term public acceptability of nuclear power. People are terrified of radiation and whilst this is a totally illogical fear, given the risks they readily accept, it is nevertheless likely to remain a major factor which will stubbornly resist all attempts to persuade them otherwise.

Widespread public acceptance is a sine qua non for the success of nuclear power. It is important because without it success cannot be guaranteed, except in the short term. In the longer term a nuclear dependency which was unpopular might require both the resort to undemocratic measures and produce major public health hazards. Nuclear power needs, more than anything else, an efficient and well-run institutional framework. It has to be well run with the highest standards of safety and management otherwise it could easily become a very dangerous industry. It is very fragile and requires a highly organised society. At the same time many countries have already passed the point of no return along the road to nuclear dependency. Once the nuclear commitment is made on any significant scale it is probably irreversible without very real risks to public safety, but these risks are not explicit enough to prevent attempts at abandonment being made.

The need for long-term public acceptability without resort to undemocratic methods is because all the major component parts of the nuclear fuel cycle involve long-term commitments that are difficult to reverse. For example, the problems range from guarding uranium mining tailings to the storage of high-level waste. The siting decision is also going to be difficult to reverse. Once a site goes nuclear it is likely to prove impossible to restore it to a natural pre-nuclear condition on any short historical timescale; the principal exceptions will be sites with small research reactors. The practicality of decommissioning and dismantlement of many civilian reactor sites looks increasingly dubious with both significant costs and health risks.

It is argued therefore that nuclear sites have to be selected which have the highest probability of remaining publicly acceptable not merely when the decision is made but also for the next few hundred years under very different socioeconomic conditions. This is required to avoid possibly later public pressure for the closure of sites for reasons of safety or health, perhaps after 2000 AD, when the medical effects of low level radiation releases are better understood or because accidents with large health consequences have occurred elsewhere in the world. It is suggested that this need for long-term public acceptability far outweighs the current preoccupation with marginal environmental and economic cost considerations. It is also important to realise that for a country the size of Britain or even the US the number of nuclear sites needed for 100 per cent nuclear electricity could be far smaller than many people imagine. In the US Burwell et al. (1979) argue for a policy of simply adding reactors to existing sites. In the UK it is not difficult to imagine that between 5 and 10 large sites could be all that is required. The question is really whether these sites should be based on existing or new locations.

The CEGB might well argue against a set of new locations on several grounds, not least because of cost, convenience, and the difficulties of abandoning existing sites. They would argue that they could never pursue a siting policy designed to maximise public acceptance of an asymptotic nuclear future because of their concern for economic generation and their statutory need to have due regard for the environment. They might even argue that their skill at interpreting public wishes is such that the existing sites are already those which are maximally acceptable to the public at large; certainly there is no real evidence that they are not. The counter-arguments are simply that the CEGB are not required to develop only the cheapest means of production, nor must they avoid all environmental damage. As for measuring public acceptability, whilst no studies have been performed to assess what is involved here, common sense dictates that most people living close to nuclear plant would prefer not to, although such preferences tend to be hidden by a large degree of apathy and are modified by CEGB propaganda of the form 'accidents will not happen here'.

If a long-term time horizon is adopted, one that looks beyond the life cycle of individual reactors, then the national interest might well be better served by a siting policy that takes into account the immense risks associated with current siting practices, and explicitly seeks to minimise these risks by the greater use of geographical isolation as a means of buying public acceptance for an all-nuclear future. The costs involved are largely economic (longer transmission lines, the need for additional capacity to make up transmission losses) and those of amenity, but these are perhaps a small price to pay to ensure the long-term public acceptance of what are in fact immense capital investments – most of which have yet to be made. In so far as there is a link between locational decisions and risk cognition and public perceptions of nuclear safety, then the time to act is now whilst the long term commitment to existing sites is not as great as it is probably destined to become. Remote siting is therefore a means of ensuring public acceptance of an energy source on which the future depends, but one with an uncertain image and one with highly emotive dangers that are not yet fully understood.

Emergency planning

The problem now is to design a measure of what are unrevealed public preferences about what makes nuclear sites acceptable or not. One way of developing such measures is to start with a consideration of existing emergency plans and public responses to both real and hypothetical nuclear emergencies as a means of identifying some of the geographical distance factors that may be involved.

It is well known that the early effects of a severe accident can be reduced by an emergency response. Aldrich *et al.* (1982) observe:

> Evacuation before containment breach within 2 miles, after release within 10 miles, and sheltering from 10 to 25 miles appear to be a particularly effective response strategy (pp. 2-104)

although Table 8.3 suggests that somewhat larger distances might be needed for a worst-case accident. Again it is noted that safety improvements and better accident consequence models have now restricted the prospects of large releases to accidents which are beyond the design basis of nuclear plant. Nevertheless public perceptions of danger or risk are not so well-informed, and can be profoundly influenced by the nature of existing emergency plans. This reflects the adage 'if it were safe then why is there an emergency plan?' and it is very difficult to provide a convincing answer.

Prior to Three Mile Island, US plant licensees were required to develop an off-site emergency preparedness plan covering the 2- to 3-mile low population zone around the reactor. Major accidents involving radiation releases that affect larger areas were thought to be highly unlikely (NRC, 1975). However, Three Mile Island drew attention to the fact that many of the emergency plans had not received full approval and also that considerable uncertainty existed with respect to the frequency of accidents.

In 1980 the NRC issued revised regulations which required all licensees to augment on-site plans by devising with state and local governments offsite plans for two emergency planning zones (EPZs); a 10-mile plume exposure pathway zone and a 50-mile ingestion exposure pathway zone. The full range of protective action was only required within the 10-mile EPZ because it was stated that the '. . . probability of large doses (of radiation) drops off substantially at about ten miles from the reactor' (NRC, 1978; pp. 1-37). The legislation also requires that there be an acoustic warning and information message system that can disseminated in the 10-mile EPZ within 15 minutes. There should be 100 per cent notification of people within 5 miles and 100 per cent within the plume exposure EPZ (a downwind $22\frac{1}{2}$ degree sector) in 45 minutes (NRC, 1980).

These revised US emergency planning arrangements are geographically far more extensive than current UK emergency plans which correspond fairly well to the pre-Three Mile Island type of US plan. According to HSE (1982) there are plans for the distribution of potassium iodate tablets and for the temporary evacuation of people from within 2 miles, and for restrictions on the distribution of milk and foodstuffs out to perhaps 25 miles. The basis of these plans is 'The dba (design basis accident) which leads to the largest off-site release' (HSE, 1982, p. 4). No attention is given to accidents beyond the design basis because 'The likelihood of such releases is so remote that further design safety measures to reduce the chance of occurrence of more extensive emergency plans are unlikely to be justified' (pp. 4-5). The US NRC would certainly disagree. This sort of perspective is one that normally characterises the nuclear industry rather than an independent public agency charged with public safety.

The logic behind this statement may well be that geographically more extensive emergency plans cannot be made for the UK because of high population densities. However, this may merely indicate that the wrong sites have been selected. Table 8.6 shows

TABLE 8.6 *Populations living within emergency planning zones in 1981*

| | Population in thousands | | |
| | within 3 | within 5 | within 10 |
Sites	miles[a]	miles[b]	miles[c]
UKAEA			
Chapelcross	12.7	15.5	28.9
Dounreay	.5	.6	11.2
Harwell	10.6	43.0	137.8
Windscale	4.9	13.7	56.2
Winfrith	8.0	12.5	83.7
Remote sites			
Berkeley	10.9	25.3	130.8
Bradwell	8.2	11.4	143.6
Dungeness	.8	8.7	18.7
Hinkley Point	1.3	6.2	82.1
Hunterston	6.8	10.7	91.7
Oldbury	4.3	35.5	190.2
Sizewell	6.9	11.0	31.5
Trawsfynydd	8.7	18.2	80.0
Wylfa	2.7	4.4	28.1
Relaxed sites			
Hartlepool	26.4	164.9	443.1
Heysham	21.0	74.1	151.2
Torness	.8	5.7	10.1
Druridge Bay	7.9	47.1	145.0

Notes: a maximum UK evacuation zone radius.
b approximate average US low population zone radius.
c post 1980 US evacuation zone radius.

the populations within 3, 5, and 10 miles of UK sites. It is very clear from these figures that large scale evacuations over US sized emergency planning zones would not be practicable for many UK sites on any kind of realistic timescale. The counter-argument is simply that similar problems also exist in the US, where it is estimated that 30 million people live within 25 miles of a nuclear plant and 85 per cent of US reactors are sited within 60 miles of a metropolitan area (Policy Research Associates, 1977). The equivalent UK population within 25 miles is 11.2 million. It would seem that the same 'judgement' has been made very differently on both sides of the Atlantic.

The differences in the geographical extent of evacuation zones is also a reflection of the US emphasis on handling accidents more severe than the design can cope with, the class 9 accident, for which siting and more extensive emergency planning measures may have a major effect on health consequences. In fact it seems that in the US the public health risks associated with design basis accidents (accidents for which the reactor's safety systems are designed to handle) are absolutely minuscule compared with the consequences of class 9 accidents. Indeed as long ago as 1975 the NRC established that class 9 accidents dominate any risk assessment of nuclear power (NRC, 1975). Whether or not any attention is given to class 9 accidents is largely a function of whether you believe the frequency estimates that have been made by reactor manufacturers. There are two difficulties in assessing the likelihood of a class 9 accident: (1) the estimated frequencies of occurrence are uncertain; and (2) the effects of factors not included; such as operator error, software errors, terrorist attacks, sabotage, and deliberate errors; are not included in the probability assessments yet they all have finite probabilities which would appear to be far larger than the risk of purely engineering malfunctions.

In the UK, the government agency concerned with emergency planning ignores the prospect of class 9 accidents and the small geographical extent of the UK emergency plans reflects a concentration on design basis accidents. The Health and Safety Executive state:

> Nevertheless the emergency plans drawn up for reference accidents provide a basic response to any emergency at a nuclear plant and are capable of extension should the need arise (HSE, 1982, p. 5).

This is hardly honest since it is extremely unlikely that either existing plans for warning populations at risk or for temporary evacuation or for potassium iodate tablet distribution would be

283

effective over the sorts of geographical areas assumed in US emergency plans. Whilst they do provide a basic response, they would probably not have much effect on the health consequences of a class 9 accident, while being increasingly unnecessary for less severe accidents.

The geographically far more extensive nature of the US emergency planning preparations have been criticised from two viewpoints. First, the pronuclear lobby has complained that there is no radiological justification for the size of the emergency planning zones since class 9 accidents will not happen. As Olds (1981) explains, 'The NRC appear to be saying that it intends to codify safety limits which many responsible critics believe already have exceeded reason' (p. 51). In other words, that the preponderance of the available evidence suggests that offsite releases from even the most severe of the nuclear accidents would have almost negligible health consequences. Additionally, other critics complain about the lack of a safety objective that such evacuation plans are meant to achieve. This sort of criticism is only natural from a nuclear industry which is trying to improve its public image. Evacuation plans and emergency measures that cover large areas may be seen to imply that nuclear power is very different from other industries and is potentially more hazardous. Of course everyone knows this is quite true, but in the propaganda battle for public support, or apathy, it is best not to be so visibly explicit about preparations for happenings which have a seemingly vanishingly small chance of ever occurring – that is according to those who make the forecasts. Quite simply emergency plans are bad publicity for the nuclear industry because of the problem of explaining their existence for plant which is, of course, completely safe and does not need it!

Another view about emergency plans is that to be effective they should take into account human behaviour. Johnson and Zeigler (1984) point out that even if it is never intended to order an evacuation extending more than 10 miles from a nuclear plant, the evacuation planning region must extend even further afield to allow for, and conform with, the behavioural intent of the population. It is exceptionally naive to believe that a population when warned of an invisible radiation hazard in their vicinity will follow the advice that originates from the same source as the radiation hazard. Should a radiation release occur at a UK nuclear station, it would be the station manager who would be held responsible, he is also the person in charge of the emergency measures to protect the civilian population. People would probably either panic or refuse to believe official assurances of well-being; indeed in the UK they might anticipate not being told anything in case the subsequent panic produced more casualties than the reactor accident. If people are informed that an accident is in progress then it is most unlikely that they will either stay put or naively conform to the emergency plan.

Behavioural responses to emergencies at nuclear plant

Three Mile Island provided a so-far unique opportunity to observe and study public responses to an actual emergency, and a number of studies have been published about behavioural responses; see Flynn (1981). These studies constitute virtually the total basis for current knowledge about human behaviour in nuclear emergencies. At Three Mile Island pre-school children and pregnant women living within 5 miles were advised to evacuate, others within 10 miles were advised to take shelter indoors. This advice should have resulted in 2500 people leaving the area; instead an estimated 144,000 within 15 miles left (Flynn, 1979) and thousands more beyond this distance also evacuated (Zeigler, Brunn and Johnson, 1981). Johnson and Zeigler (1984) note that in no other disaster has the ordered evacuation of so few resulted in the evacuation of so many. This has been termed an 'evacuation shadow phenomenon' (Zeigler, Brunn, and Johnson, 1981) or as 'spontaneous evacuation' (Chenault, Hilbert, and Reichlin, 1979). The vast difference between the observed and expected numbers of evacuees can be related to the immense public fear of radiation. So it can be argued that radiological emergency plans should be based on an understanding of the potential behaviour of the population at risk during a nuclear accident. Zeigler, Brunn and Johnson (1981) write:

> Although the evacuation-shadow phenomenon may be a minor consideration in evacuation planning for natural hazards, the impact of the phenomenon needs to be emphasised in planning for future nuclear accidents precisely because delineation of the geographical scope of an invisible danger such as ionizing radiation is difficult for public officials and private citizens to determine (p. 7).

Johnson and Zeigler (1984) surveyed 2595 households around the Shoreham plant on Long Island to ascertain likely evacuation behaviour. They estimate that approximately 49 per cent of the population of Long Island would evacuate, compared with the 3.1 per cent that live within the 10-mile emergency planning zone; or some 430,000 families compared with an expected total of 31,000. Evacuation intentions were found to vary with distance and direction. However, the evacuation distance decay curves did not register a marked discontinuity until approximately 25 miles from the plant and even then the relationship with distance was fairly poor; a correlation of −.52 between per cent of population intending to evacuate and distance. This would indicate the perceived risks far exceed the actual risks when viewed in a geographical domain. Johnson and Zeigler (1984) argue that an expanded evacuation zone would approximate more precisely the zone of behavioural intentions to evacuate, although this might further increase the zone of spontaneous evacuation.

It would seem that radiological emergencies are likely to prompt far more extreme behaviour quite different from that characteristic of other types of crises. So it can be argued that the intuitive 'folk wisdom' perception of risk may be a far more accurate and useful operational measure than actuarial statistics on which to base emergency planning measures. It might be expected that evacuation shadow phenomenon will extend to at least 25 miles beyond a plant, and probably much further. It is simply unrealistic to assume that individuals will follow governmental directives and not leave their homes even though they are located far beyond the specified evacuation zone.

Ideally these evacuation intentions could be used as a proxy for a risk cognition function and as such it should be incorporated into siting practices. The problem with this sort of approach is that the geographical distances needed to correspond to behavioural intent are very large. The median evacuation distance for evacuees from Three Mile Island was 85 to 100 miles and the zone of perceived safety had a radius of at least 45 miles (Zeigler, Brunn, and Johnson, 1981). Whilst there was a strong directional bias towards mountains upwind of the plant, the distance decay effects were less steep than anticipated. Additionally, persons living farther from the plant tended to flee to the more distant locations than did individuals living close to it.

It is quite apparent that people are absolutely terrified of being exposed to radiation from nuclear plant, irrespective of the quantity. They cannot see it or sense it. This is not simply a matter of education or mass propaganda about natural radiation levels etc.; because the fear seems to be unscientific and irrational. People also seem to strongly discriminate between natural and man-made radiation, and between nuclear industry related radiation exposure and that resulting from other beneficial exposures (i.e. medical). A population irradiated as a result of a reactor accident might well become very angry and highly annoyed out of all proportion to the likely health effects. This fear factor is something that is closely related to the acceptability of nuclear power and it really needs to be incorporated into siting practices.

Distance-related risk cognition and hazard perceptions

Geographers have known for a long time that people's behaviour in space often reflects cognitive or perceived distances about the spatial separation of objects that either cannot be seen directly or those which can (Gatrell, 1983). People may well hold perceptions about their distances from nuclear plant but at the same time have quite different cognitions as to what may be a safe distance from the unsensed radiation hazards under normal and accident conditions. Three Mile Island produced a clear link between residents' proximity to nuclear plant and their propensity to evacuate, but very little is known about whether their cognitions of risk also vary with distance (Cutter and Barnes, 1982). Yet there is a need to understand the spatial correlates of risk cognition as an input into improved emergency planning measures and siting. Nuclear power has to be acceptable to society and this implies that the risk cognition of those sections of society living near to nuclear sites has an important role to play in the locational decision, irrespective of whatever the 'real' risk might be. In a way this is an argument for a reversal of the current establishment view that the risk cognitions of experts can be used to determine publicly acceptable locations.

Cutter (1984) reports the results of a more detailed survey of Three Mile Island residents and found that, whilst there was a relationship, risk cognition was only poorly located to distances from Three Mile Island. The decision to evacuate was also influenced by the actions of friends and neighbours. It is apparent that attitudes and actual behaviour need not correspond during times of stress. Thus it would seem that whilst risk cognition is not strongly dependent on distance, actual behaviour does have a very strong spatial component. It seems that some sections of the population will want to avoid the possibility of nuclear accidents regardless of the technical basis of the threats presented. If informed of a possible danger then some people will evacuate regardless. As Sorenson (1984) notes 'All people, public and experts alike, will be confronted by uncertainty in a radiological emergency' (p. 274), and because of the complexity of nuclear technologies, and because public experience with accidents is limited, it will be difficult to issue warnings that reflect appropriate levels of risk.

The problem is that it is difficult either to identify useful distance dependent functions or a safe distance. Diggory (1956) in an early study of health hazards found that individuals farther away overestimated the extent of a health threat. But perhaps when nuclear power is involved, it would be more sensible to

287

exaggerate and overestimate distances rather than grossly under-estimate them.

Specification of a behaviouralist realistic nuclear site weighting function

It would seem therefore to be both necessary and exceptionally prudent to try and incorporate into a single evaluation measure individual perceptions of nuclear risks, their behavioural intent during accidents, and those more objective measures of the 'real' health risks produced by various computer models of reactor accident consequences. Another way of looking at this task might be to see it as involving the combination of the subjective perceptions of danger held by the public with those more objective assessments developed by scientists. This approach assumes that current attempts to bring about convergence, by altering public perceptions via education and propaganda to bring them in line with those of the nuclear industry, will not succeed. It also assumes that to be assured of long-term success nuclear power has to be regarded as both necessary and acceptable by the general public and that the locational variable is important as a major factor that may influence public acceptability. Of course all these assumptions can be criticised on the grounds that there is already popular support for nuclear power, at least in Britain, so additional attention given to 'buying' public acceptance is not needed. The inherent dangers of this argument have already been discussed and may in the long term lead to conditions which are simply unacceptable in a democratic state.

The ideal approach suggested here is to try and combine the objective and subjective aspects of the acceptability problem. The resulting weights are to be expressed as some function of geographical distance. The function cannot be based solely on the perceived relationship between 'danger' or 'safety' and distance from a reactor because of the problems of identifying a generally acceptable and wholly representative set of weights; there would be differences due to social class and individual attitudes to nuclear power as well as temporal changes which could not be predicted. At the same time the function cannot solely be a scientific measure of 'individual risks' from a class 9 accident, because these estimates are highly uncertain and conditional on all manner of assumptions. People would not know either what they mean or how to respond to them. The cigarette-smoking and life-shortening analogies of Chapter 1 are distinctly unhelpful.

The solution developed here is to combine the two approaches in the following way. The NRC (1982) provide a summary of a range of source terms (i.e. reactor accident radiation releases) and their health consequences have been expressed as individual risks should the relevant accident occur. The resulting patterns of risk have been plotted against distance, and all but one of the curves have a smooth negative exponential form. The coordinates of the highest points of any of these curves can be identified and used as a numerical measure of the relationship between distance and risk. Latent cancer fatalities rather than early deaths are used because it is viewed as a better measure of the total health impact of a reactor accident, on the grounds that the public are not likely to put much emphasis on the rather fine distinction between a prompt death and a delayed death, even if the period of delay is 10 to 20 years.

An alternative source of distance-risk relationships is given by Kelly and Clarke (1982). They provide two distance plots that represent individual risks summed over four release categories, based on two different source term assumptions; termed here as 'high' which originated from the manufacturers of the proposed Sizewell PWR reactor and 'low' which are CEGB modifications to take into account the latest information. These various risk functions are shown in Table 8.7.

The most obvious difference is the five orders of magnitude smaller individual risks from a US-designed reactor in the UK as distinct from a US-designed reactor in the US. This reflects major differences in source terms. Obviously there is little point in using any of these weights in a simple site rating function. Apart from

TABLE 8.7 *Relationships between individual latent cancer risks and distance from reactor*

Distance miles	NRC (1982) Death probability	Kelly and Clarke (1982) death probability High	Low
2	2.0×10^{-4}	8.4×10^{-11}	1.5×10^{-11}
5	2.2×10^{-5}	4.1×10^{-11}	9.7×10^{-12}
10	2.7×10^{-6}	$.7 \times 10^{-11}$	6.1×10^{-12}
15	2.2×10^{-6}	9.0×10^{-12}	4.1×10^{-12}
20	2.0×10^{-6}	7.7×10^{-12}	2.4×10^{-12}
25	1.5×10^{-7}	6.8×10^{-12}	2.9×10^{-12}
30	1.3×10^{-7}	6.1×10^{-12}	$.2 \times 10^{-12}$
35	1.2×10^{-7}	5.1×10^{-12}	9.4×10^{-13}
40	1.1×10^{-7}	4.1×10^{-12}	8.8×10^{-13}
45	1.1×10^{-7}	3.5×10^{-12}	8.1×10^{-13}
50	1.1×10^{-7}	2.8×10^{-12}	7.4×10^{-13}

the slight dilemma as to which set of weights to use, there is the problem that public perceptions and cognitions of risks from reactor accidents will not be scaled in the same way. It is most unlikely that people's perceptions of hazard would diminish so rapidly with distance – for example, by about 1000 times over a 50 mile distance. A much flatter distance decay curve would probably be more appropriate with existing work on behaviour patterns under evacuation stress as well as reflect the fact that people's perceptual or mental images are often quite different from physical reality and yet it is these images that largely determine patterns of behaviour.

Two transformations are suggested as being necessary to bring the weights in Table 8.7 into line with what may be considered as being realistic measures of public perceptions of risk. First, a logarithmic transformation is used to reduce the range of the variation of risks with distance. This is quite subjective and is justified on the grounds that the resulting flatter decay curves appear to be more intuitively plausible. It should also be pointed out that this transformation is often used to present the risks in Table 8.7 to decision-makers, and thus it would appear to be a useful metric for the perception of nuclear risks. Second, a smoothed negative exponential function is fitted to the revised data to describe the reduction of log risks with distance. The justification is threefold: (1) the curves are already of this form; (2) the smoothed values might be considered as a means of correcting for interpolation errors in constructing Table 8.7 from plotted functions; and (3) there is an extensive geographical literature to suggest that people's spatial behaviour follows a family of exponential functions (Wilson, 1970), and also aggregate perceptions or cognitions of distances for many objects follow a negative exponential function – see Lundberg (1973), Walmsley (1978); even though at the individual level there may be no such simple geometry.

Table 8.8 shows the revised sets of weights. Exponential functions of the form

$$r = a \times \exp(-b \times d)$$

where
r is the smoothed risk factor,
d is the distance from the reactor,
and a and b are parameter estimates selected to minimise the sum of squares using a nonlinear optimisation procedure –
are fitted to the data (Table 8.7) and very good levels of fit obtained. It is interesting to observe that the NRC (1982)-based

290

TABLE 8.8 *Smoothed representation of the relationship between distance and risk*

Distance miles	NRC (1982)	Kelly and Clarke (1982) High	Low
2	3.771	3.513	2.991
5	3.412	3.407	2.878
10	2.888	3.237	2.699
15	2.444	3.076	2.531
20	2.068	2.923	2.373
25	1.751	2.778	2.226
30	1.482	2.639	2.087
35	1.254	2.508	1.957
40	1.061	2.383	1.835
45	.898	2.265	1.721
50	.760	2.152	1.614
Mean absolute errors	.19	.10	.09
Parameter a	4.0312	3.5852	3.0693
Parameter b	−.0333	−.0102	−.0128

weights have a much sharper distance decay effect than the Kelly and Clarke (1982) functions.

The next problem is to decide how best to apply these weights and over what geographical range. The answers should reflect behavioural considerations as well as more objective factors. The NRC (1982) curves imply that the SST1 release would have its early deaths restricted to within 20 miles, latent cancers to about 50 miles, and land evacuation to less than 100 miles; see also Table 8.3. It can be argued that the US studies of potential evacuation behaviour indicate that at least a distance of 25 to 30 miles should be used. In the UK this would coincide with the 25-mile emergency sector used for the control of food and water. Additionally, it may be argued that the outer limit should coincide with the 50 mile emergency planning zone used in the US for the same purpose. The justification would be that people when warned about radioactive contamination of food and water, even if the levels were low, might well regard this as an unsafe region.

A final problem concerns how to compute the site index; whether to use either radial or sector populations or both. The problem with the use of radial distance bands is that the averaging of high and low density areas can produce misleading representations of demographic characteristics. This is a particular problem in the UK where the predominant preference for coastal sites offers the benefit of several zero population sectors, and these can be used to compensate for high density areas on land. Aldrich *et al.* (1982) examined 91 nuclear power sites in the US and found that the consequences of reactor accidents were largely determined by average population densities of the entire exposed population while peak results were most affected by the distance to and size of the exposed population centres. Sectors are very good at detecting population concentrations, although there is often a need to use a rotation procedure and to use more than one sector especially for periurban sites within close proximity to two or more population centres. Another consideration might be that a behavioural view of reactor accidents would not place too much emphasis on the actual physical realities of the atmospheric diffusion of a radioactive plume. It might be neat to think in terms of $22\frac{1}{2}$ degree or 30 degree sectors as a reflection of short-term atmospheric dispersal, but the residents would be unaware of the finer details of the atmospheric dispersal mechanisms and may not even care about which way the wind was blowing and whether or not it was steady or veering. The wind might well be expected to veer rather sharply especially in the coastal regions of the UK and thus confound the more naive assumptions of restricted single sector patterns of contamination. Furthermore, public safety should not have to depend on the assumption that at the precise moment when all engineered safety measures fail, the wind direction and meteorology happen to be favourable.

So it is suggested that a useful measure of the population at risk in the region of a reactor can be obtained by abandoning all tenuous links with atmospheric dispersal mechanisms and using a number of sectors carefully chosen to identify the maximum population in each of several distance bands. The problem remains, however, of how many sectors and of what angular width. Pragmatic considerations of computational convenience and population geographic aspects suggest that an angular width of 5 degrees would be a suitable choice. The use of a small width overcomes the need for a rotation procedure. Now it might be thought that the use of 5-degree sectors would result in data estimation errors, see Chapter 7. However, the search for maximum population sectors will tend to ensure that any errors are over-estimates rather than under-estimates. The final decision concerns the number of sectors that are needed to cope with all situations. Clearly 6 would be too few to allow for the best representation of population concentrations, whilst 36 might well be regarded as too many. Sufficient small sectors are needed to provide an adequate representation of both the extremes of a uniformly distributed population pattern and one with several localised concentrations at different compass orientations. In the end a decision was made in favour of 24 5 degree sectors without any restrictions on adjacency.

The site index, S, can be computed as

$$S = \sum_{j}^{11} W_j \left(\sum_{i}^{24} P_{ij} \right)$$

where

P_{ij} is the population in the ith 5 degree sector for distance band j the populations being arranged so that they are ranked into descending order of size so that P_{ij} is the ith largest of the 5 degree sectors for distance band j; and
W_j is a set of weights from Table 8.8.
The 11 distance bands are also shown in Table 8.8.

A final decision concerns the choice of weighting factors. The author's preference is for those based on NRC (1982) on the grounds that the smoothed distance decay functions show the greatest fall-off with distance. In any event it is not thought that the choice of the weights is critical, and an analysis based on the low Kelly and Clarke (1982) weights is also performed. Similarly, the selection of a 50-mile distance cut-off has been tested by a comparison with a 30-mile cut-off; the patterns are in general terms broadly the same.

The idea then is to apply this siting index to all possible 1km grid-squares within 30 or 50 miles of a populated 1km grid-square and then to rank these values into ascending order. Maps and tables can be computed for the lowest 50,000 or 100,000 or so grid-squares on the grounds that these represent optimally acceptable locations for nuclear power stations although of course within these regions the technical requirements discussed in Chapter 2 would have to be applied.

Optimally acceptable areas for nuclear power developments

The idea then is to compute site acceptability values for all 1km grid-squares in Britain and then rank them according to their scores. Areas with the lowest scores might be considered to be the most publicly acceptable, whilst areas with the highest scores might be considered the least acceptable. To avoid attaching too

293

much significance to particular values of the indices, attention is restricted to the lowest 5000, 10,000, 20,000, 50,000, 75,000, and 100,000 grid-squares. Obviously many of the locations with the lowest values will be physically unsuitable for nuclear power stations. The purpose of this exercise is merely to indicate those general regions of the country which are the most remote and thus most likely to be acceptable as a means of minimising public health risks.

The first item of interest concerns the effects of using different weights and distance summations. It would appear that there is a

TABLE 8.9 *Distribution of 'best' sites by county for the Kelly and Clarke 'high' weights: percentage of total county area for different ranking thresholds*

Region	Sites ('000s) with lowest scores					
County	5	10	20	50	75	100
Scotland						
Borders	0	0	0	25	66	74
Central	0	0	0	9	19	24
Dumfries & Galloway	0	0	0	32	98	100
Fife	0	0	0	0	0	0
Grampian	0	0	0	46	69	100
Highland	14	28	53	0	100	100
Lothian	0	0	0	0	5	8
Strathclyde	3	8	14	37	50	60
Tayside	0	0	2	25	38	72
Islands	14	21	79	93	93	93
East Anglia						
Cambridgeshire	0	0	0	0	0	1
Norfolk	0	0	0	0	1	51
Suffolk	0	0	0	0	0	5
East Midlands						
Derbyshire	0	0	0	0	0	0
Leicestershire	0	0	0	0	0	0
Lincolnshire	0	0	0	0	3	31
Northamptonshire	0	0	0	0	0	0
Nottinghamshire	0	0	0	0	0	0
North						
Cleveland	0	0	0	0	0	0
Cumbria	0	0	0	0	41	93
Durham	0	0	0	0	0	6
Northumberland	0	0	0	9	39	49
Tyne & Wear	0	0	0	0	0	0
North West						
Cheshire	0	0	0	0	0	0
Greater Manchester	0	0	0	0	0	0

TABLE 8.9 *cont.*

Region	Sites ('000s) with lowest scores					
County	5	10	20	50	75	100
Lancashire	0	0	0	0	0	0
Merseyside	0	0	0	0	0	0
South East						
Bedfordshire	0	0	0	0	0	0
Berkshire	0	0	0	0	0	0
Buckinghamshire	0	0	0	0	0	0
East Sussex	0	0	0	0	0	0
Essex	0	0	0	0	0	0
London	0	0	0	0	0	0
Hampshire	0	0	0	0	0	0
Hertfordshire	0	0	0	0	0	0
Isle of Wight	0	0	0	0	0	0
Kent	0	0	0	0	0	2
Oxfordshire	0	0	0	0	0	0
Surrey	0	0	0	0	0	0
West Sussex	0	0	0	0	0	0
South West						
Avon	0	0	0	0	0	0
Cornwall	0	0	0	0	49	100
Devon	0	0	0	0	16	43
Dorset	0	0	0	0	0	26
Gloucestershire	0	0	0	0	0	0
Somerset	0	0	0	0	0	15
Wiltshire	0	0	0	0	0	0
West Midlands						
Hereford & Worcester	0	0	0	0	4	13
Salop	0	0	0	0	6	30
Staffordshire	0	0	0	0	0	0
Warwickshire	0	0	0	0	0	0
West Midlands	0	0	0	0	0	0
Yorkshire & Humberside						
Humberside	0	0	0	0	0	0
North Yorkshire	0	0	0	0	0	10
South Yorkshire	0	0	0	0	0	0
West Yorkshire	0	0	0	0	0	0
Wales						
Clwyd	0	0	0	0	0	18
Dyfed	0	0	0	29	60	75
Gwent	0	0	0	0	0	0
Gwynedd	0	0	0	29	91	0
Mid Glamorgan	0	0	0	0	0	0
Powys	0	0	0	26	63	72
South Glamorgan	0	0	0	0	0	0

major difference between 30- and 50-mile summations but that
the choice of weights is far less important. The preference here is
for the use of 50-mile weights since this is better reflection of the
perceived areas at risk. In Table 8.9 the areas selected by the six
different rankings are reported by county based on the CEGB or
Kelly and Clarke (1982) 'high' weights in Table 8.8.

The ranking thresholds have a marked effect. For example, if
the best 50,000 grid-squares are selected then these sites are
restricted to Scotland and Wales. If the lowest 100,000 are
considered then there are areas in England as well that would

TABLE 8.10 *Distribution of 'best' sites by county for the NRC*
weights: percentage of total county areas for
different ranking thresholds

Region County	Sites ('000s) with lowest scores					
	5	10	20	50	75	100
Scotland						
Borders	0	0	0	26	67	77
Central	0	0	0	11	21	26
Dumfries & Galloway	0	0	0	34	97	100
Fife	0	0	0	0	0	0
Grampian	0	0	0	45	71	100
Highland	14	29	55	98	100	100
Lothian	0	0	0	0	5	9
Strathclyde	3	8	15	38	51	60
Tayside	0	0	3	26	41	70
Islands	12	19	63	93	93	93
East Anglia						
Cambridgeshire	0	0	0	0	0	0
Norfolk	0	0	0	0	1	43
Suffolk	0	0	0	0	0	3
East Midlands						
Derbyshire	0	0	0	0	0	0
Leicestershire	0	0	0	0	0	0
Lincolnshire	0	0	0	0	3	31
Northamptonshire	0	0	0	0	0	0
Nottinghamshire	0	0	0	0	0	0
North						
Cleveland	0	0	0	0	0	0
Cumbria	0	0	0	0	40	93
Durham	0	0	0	0	0	10
Northumberland	0	0	0	10	41	52
Tyne & Wear	0	0	0	0	0	0
North West						
Cheshire	0	0	0	0	0	0
Greater Manchester	0	0	0	0	0	0

TABLE 8.10 *cont.*

Region County	Sites ('000s) with lowest scores					
	5	10	20	50	75	100
Lancashire	0	0	0	0	0	0
Merseyside	0	0	0	0	0	0
South East						
Bedfordshire	0	0	0	0	0	0
Berkshire	0	0	0	0	0	0
Buckinghamshire	0	0	0	0	0	0
East Sussex	0	0	0	0	0	0
Essex	0	0	0	0	0	0
London	0	0	0	0	0	0
Hampshire	0	0	0	0	0	0
Hertfordshire	0	0	0	0	0	0
Isle of Wight	0	0	0	0	0	0
Kent	0	0	0	0	0	0
Oxfordshire	0	0	0	0	0	0
Surrey	0	0	0	0	0	0
West Sussex	0	0	0	0	0	0
South West						
Avon	0	0	0	0	0	0
Cornwall	0	0	0	0	22	94
Devon	0	0	0	0	17	42
Dorset	0	0	0	0	0	25
Gloucestershire	0	0	0	0	0	0
Somerset	0	0	0	0	0	16
Wiltshire	0	0	0	0	0	0
West Midlands						
Hereford & Worcester	0	0	0	0	5	14
Salop	0	0	0	0	9	31
Staffordshire	0	0	0	0	0	0
Warwickshire	0	0	0	0	0	0
West Midlands	0	0	0	0	0	0
Yorkshire & Humberside						
Humberside	0	0	0	0	0	0
North Yorkshire	0	0	0	0	1	12
South Yorkshire	0	0	0	0	0	0
West Yorkshire	0	0	0	0	0	0
Wales						
Clwyd	0	0	0	0	0	21
Dyfed	0	0	0	25	61	77
Gwent	0	0	0	0	0	0
Gwynedd	0	0	0	29	88	0
Mid Glamorgan	0	0	0	0	0	0
Powys	0	0	0	28	65	73
South Glamorgan	0	0	0	0	0	0

FIGURE 8.1 Location of the best 5,000 1km grid-squares in Britain

FIGURE 8.2 Location of the best 10,000 1km grid-squares in Britain

FIGURE 8.3 Location of the best 20,000 1km grid-squares in Britain

FIGURE 8.4 *Location of the best 50,000 1km grid-squares in Britain*

FIGURE 8.5 *Location of the best 75,000 1km grid-squares in Britain*

100 Km

FIGURE 8.6 Location of the best 100,000 1km grid-squares in Britain

qualify. Table 8.10 gives the same range of rankings but using the NRC (US) derived weights. The overall pattern is very similar. Figures 8.1 to 8.6 show the locations of these most acceptable areas. The squareness of some of the maps is due to edge effects resulting from the ranking procedure.

Another way of examining these results is by focusing attention only on those areas which occur on the UK coastline. This is an easy way of removing the more unlikely locations, particularly those on mountain tops! Table 8.11 reports the proportion of each county's coastal grid-squares which are included in the

TABLE 8.11 *Distribution of 'best' coastal sites for the Kelly and Clarke 'high' weights: percentage of county coastal areas for different ranking thresholds*

| Region | Sites ('000s) with lowest scores | | | | | |
County	5	10	20	50	75	100
Scotland						
Borders	0	0	0	20	20	20
Central	0	0	0	0	0	0
Dumfries & Galloway	0	0	0	40	79	79
Fife	0	0	0	0	0	0
Grampian	0	0	0	2	14	54
Highland	14	31	39	58	58	58
Lothian	0	0	0	0	0	0
Strathclyde	5	12	18	34	40	42
Tayside	0	0	0	0	0	35
Islands	21	24	46	53	53	53
East Anglia						
Cambridgeshire	0	0	0	0	0	0
Norfolk	0	0	0	0	0	13
Suffolk	0	0	0	0	0	20
East Midlands						
Derbyshire	0	0	0	0	0	0
Leicestershire	0	0	0	0	0	0
Lincolnshire	0	0	0	0	11	44
Northamptonshire	0	0	0	0	0	0
Nottinghamshire	0	0	0	0	0	0
North						
Cleveland	0	0	0	0	0	0
Cumbria	0	0	0	0	10	43
Durham	0	0	0	0	0	0
Northumberland	0	0	0	4	4	6
Tyne & Wear	0	0	0	0	0	0
North West						
Cheshire	0	0	0	0	0	0
Greater Manchester	0	0	0	0	0	0

TABLE 8.11 *cont.*

Region County	*Sites ('000s) with lowest scores*					
	5	*10*	*20*	*50*	*75*	*100*
Lancashire	0	0	0	0	0	0
Merseyside	0	0	0	0	0	0
South East						
Bedfordshire	0	0	0	0	0	0
Berkshire	0	0	0	0	0	0
Buckinghamshire	0	0	0	0	0	0
East Sussex	0	0	0	0	0	0
Essex	0	0	0	0	0	0
London	0	0	0	0	0	0
Hampshire	0	0	0	0	0	0
Hertfordshire	0	0	0	0	0	0
Isle of Wight	0	0	0	0	0	0
Kent	0	0	0	0	0	6
Oxfordshire	0	0	0	0	0	0
Surrey	0	0	0	0	0	0
West Sussex	0	0	0	0	0	0
South West						
Avon	0	0	0	0	0	0
Cornwall	0	0	0	0	31	54
Devon	0	0	0	0	8	26
Dorset	0	0	0	0	0	27
Gloucestershire	0	0	0	0	0	0
Somerset	0	0	0	0	0	0
Wiltshire	0	0	0	0	0	0
West Midlands						
Hereford & Worcester	0	0	0	0	0	0
Salop	0	0	0	0	0	0
Staffordshire	0	0	0	0	0	0
Warwickshire	0	0	0	0	0	0
West Midlands	0	0	0	0	0	0
Yorkshire & Humberside						
Humberside	0	0	0	0	0	0
North Yorkshire	0	0	0	0	0	2
South Yorkshire	0	0	0	0	0	0
West Yorkshire	0	0	0	0	0	0
Wales						
Clwyd	0	0	0	0	0	0
Dyfed	0	0	0	28	44	49
Gwent	0	0	0	0	0	0
Gwynedd	0	0	0	24	53	56
Mid Glamorgan	0	0	0	0	0	0
Powys	0	0	0	0	0	0
South Glamorgan	0	0	0	0	0	0

results shown in Table 8.9. There would appear to be no shortage of sites in certain areas. It should be noted that all these locations would in fact satisfy the existing UK relaxed siting demographic criteria for nuclear power stations. Finally, it is observed that none of the existing nuclear power station sites lie within any of these optimally acceptable areas.

The purpose of this chapter was to indicate in a broad and consistent manner those parts of Britain where the CEGB or SSEB should be prospecting for sites. The CEGB and SSEB will of course conjure up all manner of excuse as to why remote siting of this kind is wrong, misleading, meaningless, and technically impossible. They will criticise the derivation of the weights and question whether or not public opinion bears any relationship to them. They may argue that the strength of the 'environmental lobby' is so great that real remote siting is not acceptable. The response to such criticisms is merely that public acceptance of nuclear power is largely a function of location. The kinds of sites that the CEGB prefer in semi-urban and non-remote semi-rural locations are already proving to be most unpopular and should any major accident ever occur, there will be pressure to close them down. Quite simply there are too many people living within the geographical range of a major reactor accident. The oil and gas industry have shown how it is possible to ship large quantities of energy from one end of this small country to the other as a matter of routine. If they can do, then why cannot the CEGB? Of course there are costs involved but these could be seen as the price that nuclear power has to pay for public acceptance. In any case it seems that bulk energy transfers of the sort envisaged here will one day become a standard feature of power generation. The old strategy of locating new facilities close to major demand centres is, or will soon be, a relic of the past when the units of generation were small enough to preclude any other form of spatial organisation.

Conclusions

The purpose of this chapter is to develop the argument that the long-term public acceptability of nuclear power is critically dependent on locational decisions made at a time when general public perceptions of risk and health hazards are under-developed. The most robust solution is not to seek to teach the public that the experts are right, nor is it purely a matter of seeking to identify minimum casualty locations with respect to a nuclear industry 'worst-case' accident, but it involves searching for maximally acceptable sites. This is important because an occurrence of a major reactor accident anywhere in the world would result in the majority of the UK population feeling themselves to be at risk. Table 8.12 gives an indication of the numbers of people living within various distances of nuclear power stations. In total 68 per cent of the population live within 50 miles of one or more nuclear power stations. So it would seem that under certain circumstances these people may believe themselves to be threatened. They may become very concerned about the poor location of existing nuclear plant whilst completely dismissing narrowly focused economic arguments of the form 'remote sites cost more and will raise electricity bills'. Likewise cost-benefit analyses in which virtually no costs are attributed to the health consequences of nuclear power can be rejected as being myopic beyond belief. In short the current planting strategy used for nuclear power stations is a gamble with public funds and public safety. The longer the current pattern continues the more difficult and the more expensive it becomes to change it. The more entrenched the attitudes the more difficulties that confront any alternative spatial arrangement.

To start this kind of debate off, a series of weights were derived to represent a combination of subjective and objective risk assessments so that the 'best' areas could be identified. The site acceptability index that was derived might be considered as a weighted risk function although it has no natural interpretation either in terms of potential casualties or in monetary terms. It is used merely to indicate a particular type of ordering that can be used to guide the siting process. It is based on a subjective interpretation of what might be called a public acceptability function, although of course further refinement is clearly necessary. Nevertheless, the sites identified in Tables 8.9 to 8.11

TABLE 8.12 *Populations at risk from nuclear power developments in Britain*

Site	1981 populations (in millions) within these distances of one or more nuclear installations			
	10 miles	20 miles	25 miles	50 miles
UKAEA only	.22	1.16	1.73	9.35
'Remote' sites	.90	5.28	7.99	26.97
'Relaxed' sites	1.49	6.91	10.24	36.23
All sites	1.63	7.72	11.27	36.24

Notes: Current sites only.
Populations are cumulative down the columns.

might be expected to threaten less than 1 million people; at least 36 times fewer than at present.

It really does not matter that the weighting functions are imperfect. The objective was merely to identify at a strategic level those areas which looked the most promising. Likewise there is no attempt at cost benefit analysis; see for example, Niehaus and Otway (1977). Such assessments are irrelevant because given present assumptions that nuclear power is completely safe and has no adverse health effects whatsoever, it is impossible to demonstrate any benefits resulting from any additional safety measures. Since we are told that British reactors do not release any harmful radiation under any credible conditions there is nothing that can be gained by using siting as a design-independent and additional safety factor. Likewise, the financial benefits of siting to maximise public acceptability in the long term cannot be costed either. On the other hand, the costs of such a strategy can be readily enumerated and may seem overwhelming. However, many of the additional costs are of a 'one-off' nature – for example, transmission facilities. If a suitable long-term view is taken – for example, 300 years rather than 25 – then these costs are trivial. If the risk of closure of the nuclear plant for political reasons were to be costed, and if these reasons were to be related to lack of public confidence in the nuclear industry and its siting practices, then it is fairly certain that the adoption of a real remote siting strategy might even be seen to save money.

Environmental and amenity objections may seem fundamental but people given a choice between the localised devastation of a few remote beauty spots compared with their own location in close proximity (within 50 miles) of nuclear plant with an increasingly suspect safety record (all local occurrences of anything different from national averages will be blamed on it), will prefer remote devastation. Such a planting strategy may in any case emerge sometime in the next century so an early move in that direction may well yield long term benefits. The most obvious locations are in remote areas near to the gas and oil pipelines. It is more than likely that in the long term nuclear electricity will be used to produce replacement gas supplies – i.e. hydrogen – when North Sea supplies are exhausted. This chapter then is a plea for a new look at the sorts of areas where more detailed site searches should be performed, bearing in mind the need to adopt a long-term perspective. The planting strategy needs to be part of a national strategic plan for energy provision in the 21st and 22nd centuries. It cannot simply emerge incrementally to meet short-term goals with no view towards either the longevity of the sites themselves, or the long-term pattern of energy supply.

9 Summary, conclusions and recommendations

The book has been concerned with the link between reactor siting policies and public safety in the UK and to some lesser extent in the US. Of particular concern has been the general lack of attention afforded siting as a safety measure in those countries that are currently still developing nuclear power (UK, France) compared with an arguably more realistic siting perspective in the US, where new nuclear power developments are not now expected until perhaps the next century. The book was written at a time when nuclear power supplied about 12 per cent of the world's electrical energy, when the number of operating stations exceeded 300, and the accumulated number of years of reactor operating experience exceeds 3100 (*Atom*, 1984, 334, p. 23). Despite this currently small level of nuclear generation it should not be forgotten that by 2000 AD and certainly by 2030 AD the situation will probably be very different with about half or perhaps even two-thirds of all electricity being nuclear powered. In many ways the 1980s represent the beginnings of a rapid increase towards eventual nuclear saturation of electricity generation at 80 to 90 per cent some time during the twenty-first century; or at least that is what at present seems to be the outcome of current intentions.

The best available scientific evidence seems to suggest that the risks of a major nuclear accident with more than 1000 deaths is of the order of one in a million per year, so even with 300 stations operating there is an approximate world-wide risk of 1 in 3333 per year. In the UK nuclear saturation will probably mean less than 100 stations so that risks of 1 in 10000 would appear to be of the right order. This is supposed to be comforting because it has been calculated that a major accident at Canvey Island (probably the most dangerous place in the UK due to a mixture of people, oil, and gas facilities) was estimated at 1 in 5000. Another way of

310

expressing these numbers is to assume (wrongly of course) that they represent frequencies of occurrence; for example, a risk of 1 in 10,000 would imply a major accident once every 10,000 years, other things being equal (which of course they are not). The correct interpretation is that these are expected to be rare events which could happen at any time. It must indeed be very reassuring to the British people that one day the risks from the operation of nuclear power stations are twice as small as those currently associated with living on Canvey Island! These risk assumptions are also, by the way, dependent on the simple assumption that reactors are only subjected to random equipment failures, and they exclude any possibility of deliberately induced accidents due to operator error, terrorist attack, or sabotage. They are excluded because no reliable probability estimates could be attached to them. A final assumption is, of course, that people will assess the numerical risks of nuclear power on the same scale as they are supposed to assess other risks associated with modern living – for example, that nuclear power is twice as safe as Canvey Island because the risks are twice as small. This may be a gross error since the effects of a major reactor accident may be fundamentally different from an explosion at an oil storage plant. In particular, there may be need for a long-term evacuation of contaminated areas and the risk of the contamination being diffused by fluvial processes and various ecosystems far beyond the immediate area of concern.

Nevertheless, the numbers can be made to look even more convincing by making the probabilities smaller, either by reassessing the accident probability frequencies (contemporary UK PWR accident frequencies seem to be four or five orders of magnitude smaller than US equivalents) resulting from a better theoretical understanding of accidents, or by converting the crude probabilities into individual probabilities. The latter would show that the chance of a person dying as a result of living near a nuclear power station is about 1 in 100,000. If this were to be applied to the entire population of Britain it would result in about 5,300 deaths every 40-year period; not many really. The problem, however, is that of convincing the public that these sorts of statement are even approximately correct. It is a problem because of uncertainties in the calculations and the need for an empirical data base covering perhaps 10,000 or more years of reactor operation before there can be any real assessment of risks. In other words accurate risk assessments can only be done retrospectively, and nuclear power has not yet been around long enough. Which set of figures you believe at present is largely a matter of faith. If you like nuclear power than you will certainly

311

be able to find an appropriate set of numbers to confirm this belief. Likewise, if you think it is dangerous and should be abolished then it is also possible to use the same figures to support this belief.

The problem of convincing the public is a very important one. The nuclear industry knows very well that it must win and retain public acceptability for all aspects of the nuclear fuel cycle; at least whilst there is still time for public and political pressure to stop its development. The method of approach they have adopted has not yet worked; the economic advantages of cheap electricity if you discount various items (i.e. development costs, waste disposal etc.) have not yet won the day for the nuclear industry. Instead they face a steadily worsening media representation. The 'bad press' over the Sellafield leaks and the public relations disaster by NIREX (the waste disposal executive) of trying to justify the selection of a site for an intermediate level waste disposal facility under a town testify to the difficulties of the task ahead. At the same time it would appear that there is growing, totally circumstantial medical evidence that proximity to nuclear facilities is bad for your health, and more especially that of your child's; it is difficult to imagine a more emotive issue.

Given that these problems need to be overcome, how should the nuclear industry attempt to launder their image? One way is for eminent scientists to unequivocally state that the current probability assessments are tantamount to saying that nuclear power is completely safe. The message could then be spread via a good firm of PR consultants. At the same time you would of course spend more money on school-level education. Perhaps the MoD would share the cost, then at least future generations of citizens would be able to readily understand the role of nuclear weapons as a means of ensuring world peace and grasp the horrors of living in a world that did not accept nuclear power. Alternatively, you could try mass indoctrination!

The contention here is that there is a far simpler approach which involves the adoption of a siting strategy that explicitly seeks to minimise health consequences from worst-case accidents, regardless of however rare such events may seem to be according to the best estimates of the nuclear industry. Then at least nuclear power could be sold to the public under the banner that all possible steps had been taken to minimise the consequences of any radiological accident, however rare and however caused. This incidentally, is almost identical to one of the UK's HSE's nuclear safety objectives, which at present is being interpreted in a rather narrow way. If you could be certain without any uncertainty that nuclear power was completely safe then there

312

would be no need for such costly measures. The problem at present, and, one suspects, for the foreseeable future, is that no such guarantees can be made by honest people. In any case real remote siting might appear to be more uneconomic than it actually is. If the benefits of remote siting were to include a monetary evaluation of both a guaranteed future for the existing nuclear facilities without any risk of closure for political reasons and the cost of a major accident, then one imagines that a most favourable return on the additional expenditure would be assured. There is also a precedent for what may be regarded in terms of present day criteria as uneconomic, unnecessary, and non-scientific courses of action; for example, BNFL's decision to reduce the amounts of harmless radioactive waste they pump into the Irish Sea from Sellafield in deference to public pressure.

In some ways the basic public safety problem can be summarised in the form of a map. Suppose that the hazard range of a possible nuclear accident could be set at a maximum of 50 miles. Then Figure 9.1 shows the area of the UK covered by 50-mile radius circles centred on existing nuclear power station sites. The 36.24 million people (see Table 8.12) living in these areas might see themselves – rightly or wrongly is largely irrelevant – as being at some unquantified risk. This sort of perception would readily happen after the occurrence of a major reactor accident. Suppose now that a full commitment to an all-nuclear future were to be made. The revised set of circles based on the sites plotted in Figure 2.1 would result in virtual blanket coverage of the entire UK population; see Figure 9.2. The question that should be asked is whether or not it is sensible to go for a dispersal of nuclear power station sites with a consequential global at-risk population or whether or not there should be a policy of concentration. If the latter, then what proportion of the UK population should be put at risk? The optimal sites identified in Chapter 8, see Figures 8.1 to 8.6, would effectively minimise this quantity; whereas a continuation with the existing sites would seem to represent a half-way house, but one that still exposes too many people to possible plume exposure pathways. The expected response to this sort of argument is to assume that, providing reactors are designed to incorporate sufficient engineering safeguards, the likelihood of a serious accident can be discounted. Britain must be one of the few countries in the world where the safety agencies implicitly believe and trust in the absolute infallibility of reactor designs. It is a very fragile basis on which to establish a large-scale nuclear power programme.

Returning to the contents of this book, the examination of various aspects of past and present reactor siting strategies in the

313

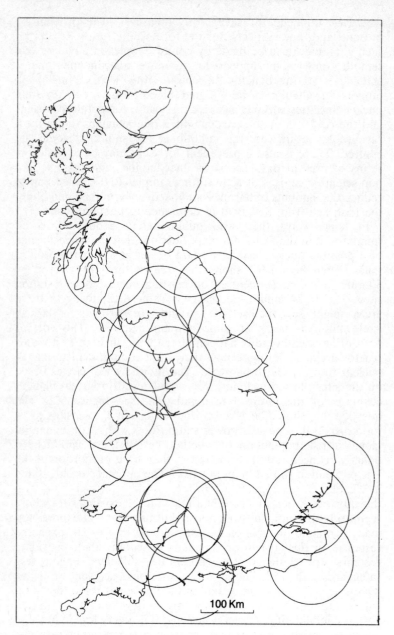

FIGURE 9.1 Areas of the UK covered by 50-mile radius circles based on current nuclear power station sites

FIGURE 9.2 *Areas of the UK covered by 50 mile radius circles based on sites identified as of possible interest as locations of nuclear power stations*

315

UK and USA leads to two basic conclusions:

(1) there has been a general failure to appreciate the extent to which location is a very important variable in measuring the safety (or danger) potential of nuclear sites; and

(2) there has been a failure to realise that location is also a critical factor that affects the level of present and future acceptability of nuclear power.

There are, of course, good reasons for this neglect: engineering convenience and short-term economic factors favour sites near to transmission lines and in areas of major demand. Additionally, current siting practices evolved in an era when siting decisions were largely secret and attracted little attention. In the US the early 1970s saw an opening up of the siting process to public scrutiny with a great increase in public opposition to nuclear plans. In the UK it seems that perhaps the Sizewell Public Inquiry (1983-5) will serve a similar function. It is certainly long overdue as previously the public's role in nuclear siting decisions has been very limited; it was largely one-way with the CEGB explaining the logic of decisions which in fact could not be reversed except under most exceptional circumstances of the sort that do not occur.

As a planning exercise, the CEGB's siting practices are outmoded and out of date. In most other areas of planning there is an evaluation of alternatives, in full, before any decisions are made. At present the CEGB make the decision first, on the basis of a partial evaluation of a site, and then justify it later when additional information is available. As Openshaw (1982b) suggested, they seem to select a site for reasons best kept to themselves, and then later publicly justify it for quite different reasons. Now it might well be thought that it is simply not sensible to go about locating multi-billion-pound power complexes with a longevity that will probably be measurable eventually on geological timescales, without the most thorough and systematic examination and evaluation of all possible alternative sites throughout the entire country. Such an evaluation process must be formalised, comprehensive in coverage, and performed in public in order to demonstrate that there is in fact a logical and rigorous process of site selection. Why should people, or indeed the government, believe that the current, partial, and largely invisible siting process is optimal? How, in any case, can any sites be identified without a full set of approved national siting objectives? It is not enough to answer the question 'Why here?' with the statement that the site is technically suitable and is the result of a well-rounded choice! The problem is, how was the well-rounded choice made, and were the full range of possible alternatives considered. There must be a large number (several thousand) of technically suitable sites around the coast of Britain, let alone on rivers and lakes, so why single out one site in particular? How were the comparisons made? And the answer appears to be that you first exclude 99.999 per cent of all sites and then concentrate on the remaining two.

In some ways it is not the CEGB's fault. They do their best to meet their statutory duties. The problem is that the interpretation of these statutory duties leaves them with very little lee-way. Perhaps it is time that a different government department was given the responsibility for siting issues so that the delicate balance between environment and power economics could be resolved in a nationally acceptable fashion. Certainly, the present situation is not satisfactory and will mainly serve to create more problems in the future.

The Sizewell Public Inquiry served to demonstrate major procedural problems. The general point is nicely summarised in a question by Mr Brooke, Counsel to the inquiry. He asked:

> Mr Gammon, if an unkind outside critic was listening to what you are saying, perhaps they would say something like this to the inspector: Just look at the timescale. Here we are. We are now considering the details of the evidence 6 years after the decision and the Board tell us, well, you cannot put the decision into reverse because it would cost the electricity consumer too much money. But when the Board is actually carrying out its statutory duty, it is rather hit-or-miss what each individual member of Board may or may not have remembered from the last time they considered the matter (Sizewell B Public Inquiry, 1984, p. 56, Day 249).

The last statement is a reference to the fact that the choice of Sizewell was made in 1978 without any formal evaluation of alternatives, on the grounds that the Board members knew all the details from previous applications to build nuclear power stations there in 1958, 1969 and 1973. The critic would probably also point out that the decision was made at a time when there was no reference design. Additionally, most of the information needed to evaluate the decision and fully evaluate the site (in pursuit of Section 2 and 39 duties) did not exist when the decision was made. Yet the CEGB warned that every year that the Sizewell build-decision is delayed it would cost electricity consumers millions of pounds, it would cause problems to the morale of the CEGB design staff, and it would make the next public inquiry for the next nuclear application even more protracted. It would seem that so many plans had been made before the Inspector came to make his recommendation that the Board is in fact saying that even if the decision is wrong it would cost the consumer too much to unscramble it. This is hardly a satisfactory way to run such an important enterprise, and the fault is partly due to the nature of the existing planning legislation.

The tremendous difficulties of the CEGB proving that any site is in fact the best obtainable balance is compounded by the mistaken assumption that nuclear power is publicly acceptable to such an extent that fine tuning the economics of siting and design is worthwhile. There is no evidence to support this feeling of consensus; indeed, one suspects that the majority of the public are apathetic and do not yet have an opinion about the desirability of nuclear power. Indeed, if anything there would be a consensus against it. Many people are becoming highly sensitivised to radiation and are absolutely terrified, in an irrational way. It is probably only the complexity and mysticism associated with most aspects of nuclear power that has prevented a polarisation of views and the entry of political parties. One imagines that this honeymoon era will not last much longer. Public apathy and lack of understanding offers no basis on which to build a multi-billion-pound nuclear future with public funds. The CEGB proudly claim that they have noticed a high degree of trust and acceptance of nuclear power from people who live next to their nuclear plant. Such trust is rewarded with the construction of additional reactors; at least in France they receive discounts on their electricity bills as well! However, it may be observed that this feeling of trust is largely a reflection of repeated CEGB statements that no harm will or can be caused by their plant. It is a rather fragile peace and the guarantee is by no means absolute. One imagines that if the local residents could actually comprehend the vast amounts of radiation stored there, and the effects on them were significant amounts to be released, then they might well be less complacent. But even if there was popular alarm, there is no mechanism whereby this could be used to affect subsequent CEGB decisions if the Board members chose to ignore it.

In general terms, nuclear safety is not purely a matter of building infallible control and containment systems for reactors to be operated by equally infallible experts. The systems being controlled are not purely physical or closed systems but they are open and their safe operation depends on human institutions, and the absence of external events that it cannot handle; for example, an insane operator, a major earthquake, a small war, or sabotage. Responsibility for finding a safe site is not merely a matter of meeting demographic and related requirements set down by the NII or NRC. In the UK safety considerations which are site-dependent exist as a constraint on a quasi-optimisation exercise that has an economic evaluation criteria as the objective function. These demographic constraints are incredibly lax.

It is also necessary for the nuclear industry to be honest in their safety assessments. Why not admit that whilst nuclear power is exceptionally safe, it is inevitable that a small number of unfortunate people might die as a direct result of their nuclear operations under normal accident-free conditions. This might well be estimated as a small fraction of deaths that could be attributable to non-nuclear alternatives. However, the practice has been to deny that there ever have been any civilian casualties and to insist that none are expected. The more such absolute guarantees of safety are given, the more difficult it becomes to handle increasing amounts of contradictory circumstantial evidence of the sort that sells newspapers. The longer this process continues, the greater the gap between fact and fiction and the less public credibility the nuclear industry will have. The process is completely self-defeating – especially, as the public probably do not believe these utterances anyway. The problem the nuclear industry faces is that were they to admit to a small number of deaths, even over a long period of time, then the public may well decide that any excess deaths are to be adverted. It may be that people prefer to die from lung diseases induced by atmospheric pollution from conventional plant than from radiation-induced cancers from nuclear power. If they do, then that would be a legitimate democratic choice. Among other things it might result in the closure of particular sites and a greater urgency about seeking real remote locations. All this could happen without having to wait for a major reactor accident to occur.

Certainly there is evidence of a greater degree of realism in the US. It is increasing doubts about the safety of nuclear power and the growing fear of long-term low-level radiation effects, that have been behind recent moves to concentrate new developments only on the most remote sites. It has been noted in a previous chapter that many UK sites have demographic characteristics very similar, or worse, than those US sites which will probably cease nuclear operations once existing facilities reach the end of their operational lives. There is a growing feeling that additional safeguards are needed, safeguards which are independent of the nuclear establishment and immune to their manipulations. The use of real remote sites is an easy way of offering the public additional insurance. It will almost certainly help overcome some of the problems of public acceptability, at

319

least for the majority of people.

In the US the need for public support is also clearly recognised. Although all the binding referenda that would have shut down existing nuclear plant have been rejected, some of the decisions were very close. Furthermore, referenda and legislation have been approved in 11 states, and this will prevent the construction of any new nuclear plant unless various prescribed conditions are met. It would seem that the success of any new nuclear projects will depend on them obtaining majority public support. At the same time a central factor for public concern is the fear of a nuclear accident with severe health effects. One consequence has been that the NRC has imposed safety backfits to many existing reactors in an attempt to allay these fears. Another is the greater emphasis on the developmen of alternative reactor technologies; including gas cooled reactor systems of the sort the British have abandoned in favour of the PWR. According to OTA (1984) current PWR designs are unlikely to be viable choices for the next era of nuclear reactor building in the US unless various outstanding concerns over cost, safety, and economics can be alleviated. Meanwhile the US nuclear industry plans to survive the current depression partly by supplying their obsolete (in terms of US criteria) plant to countries stupid enough to think they are a good buy.

In Britain it would seem that the choice of the PWR is merely a stop-gap measure prior to the dawn of commercial fast breeders. They seem to offer cheaper electricity than the home-made AGR and they can be mass-produced. Their existence is necessary both to ensure a future dependency on nuclear electricity and, possibly, to provide fuel loads for the subsequent fast breeders. It is most unlikely that any British government would sanction referenda either on regional or national lines to test the acceptability of nuclear power. It would be far too risky. Instead the British people have to trust the experts in government and the nuclear industry whose job it is to ensure their safety. Secrecy is used to hide the tremendous conflation of interests that arises when all the so-called independent agencies are all acting under government policy. This entire cosy arrangement will probably continue until a major disaster forces a fully independent inquiry on the entire regulatory, inspecting, policy making, and utility organisation process. It is an exceptionally fragile basis on which to build a major, nationally important industry. In many ways it is predicated on an immense gamble with public money and public safety. If the gamble succeeds then the country will have slightly cheaper electricity than otherwise. The use of semi-urban sites to demonstrate

320

government confidence in the safety of nuclear power will have been vindicated and the experiment with public safety declared successful; although such a decision is left to future historians to make. If it fails, then the entire nuclear infrastructure is at risk. Should there be a major accident near to an urban region the public will be immensely reassured to know that the offending organisation may be liable to prosecution. Additionally, the public may receive some compensation if they can prove in law that their death or health problems are due to a particular nuclear accident, perhaps 30 years previously. An informed public might be exceptionally adverse to this kind of risk taking when they have little to gain and much to lose.

The zero-infinity dilemma is a very real one. We should be asking how robust is the current nuclear planting strategy in relation to the political and public effects of a major accident? The problem is that there is no government department which is charged with this sort of responsibility. It is certainly true that the odds are heavily stacked against a nuclear disaster here, but the odds are far less favourable for the world as a whole. The second question, then, is how closely linked is UK public opinion to events in the rest of the world? Indeed, if one adopts this global perspective, then each new nuclear power station really has to be built to more stringent safety standards than the previous one in order to slow down the increase in total cumulative risks.

A final matter concerns what advice can be provided to both government and the nuclear industry based on the contents of this book. A number of definite statements can be made.

(1) The selection of all new nuclear power sites, and the locations of all new reactors, must be made to satisfy explicit public safety criteria which are reactor independent but site dependent with a view towards establishing the acceptability of nucler power over long historical timescales; at least one, possibly three centuries. The current practice of considering only design basis accidents and not the more severe class 9 accidents is to be deplored as it provides no satisfactory basis for demographic siting criteria.

(2) There has to be a national siting policy which has been debated by Parliament and is publicly acceptable. Such a policy would be based on a set of objectives and would follow from a detailed study of alternative locational strategies. Nuclear power cannot be allowed to locate itself in an ad hoc and piecemeal fashion with decisions being made on an individual basis. There has to be a long-term plan or strategy and a vision of what the end-state will involve.

(3) The current environmental impact and assessment processes require a complete overhaul. The Sizewell Inquiry clearly showed the absence of any formal structure as to how the CEGB were to take the Section 37 duties into account.

It is not enough to claim that all their employees have due regard to these matters throughout all their work (Sizewell B Public Inquiry, 1984, p. 99, Day 247). There has to be an explicit methodology that can be used to assess the merits of alternative sites and alternative locations and a means of including environmental costs into the economic assessments that are used. If such techniques cannot be devised, which is frankly unlikely, then Government should review how the environmental duties of Section 37 are to be fulfilled. Additionally, the present practice of giving the CEGB the sole responsibility for finding acceptable sites in relation to both their own requirements and environmental aspects should be discontinued forthwith. The US practice of relying more on reconnaissance-level material would allow a far more comprehensive evaluation to be made of alternative sites at low cost. You cannot reasonably restrict attention to two or three locations or sites in one region of the country and then pretend that a proper evaluation has been performed.

The basic principles of how to go about evaluating alternative sites are well known. Keeney (1980) states what is really very obvious to all except the CEGB:

> The first step in an energy facility siting study is to identify a reasonable number of candidate sites. The set should contain the best feasible sites for the proposed facility. However, because of the complexity of the problem it is usually not possible to pick the best among these sites without a formal analysis (p. 80).

The point to emphasise here are the words *formal analysis*. It is precisely an absence of in-depth formal analysis that characterises the current CEGB siting practice. Sizewell it seems was justified retrospectively. Of course a proper evaluation is not easy and will certainly prove expensive, but surely the size of the proposed capital investments is sufficient to justify the most elaborate identification and evaluation procedures. The siting of nuclear power stations is perhaps the most important locational decision that will be made during the remainder of the twentieth century and it is extremely important that only the best possible sites are selected. The days have long gone when such facilities could be sited by engineering and economic criteria alone. If sites are to be found that provide a balanced and rounded decision which takes into account the competing interests of the engineer, the environment, and the need for public safety – then it is only fair that the basis for the decision should be formalised and open to public scrutiny. It is not enough to claim that the need to gain planning consents is a sufficient test; because by then the decision has been made and the issue reduced to a simple 'yes' or 'no' confirmation.

(4) The existence of a current site licence at a location dating from the MAGNOX or AGR era of reactors should not be allowed to confer any special advantage when considering the suitability of a site for new developments or for re-developments. Indeed it is quite clear that many existing UK sites are exceptionally poorly located and should be abandoned rather than redeveloped. The current British practice of naively assuming that once a site is judged to be technically suitable and that nuclear developments there are environmentally acceptable, should be discontinued forthwith. The historical justification for a site may no longer be relevant. Whilst it may be nice to consider minimising the need for new greenfield sites by seeking to fully develop all existing sites (an advantage for the utility), this really has to be a decision by government. Radioactive contaminants tend to accumulate in environments and the dangers of growing radioactive pollution in areas near to large population concentrations should be identified before the potential health problems start to attract public attention. An obsequious affiliation with the short-term needs of the transmission system should not be allowed to either become or to remain the principal location determining variable. The dangers to public health from normal operation and from accidents need to be given far more attention; indeed, they should be paramount.

(5) The manner by which sites are selected and approved needs radical changes to allow a two-way input of public views and to take into account the full range of sites that may be considered feasible in terms of the few criteria that really matter. Unless and until this geographical process is democratised the future public acceptability of nuclear power will remain in doubt. The urgency and the importance of electricity developments should not be allowed to overrule the need to evaluate all alternatives, not just a few. It is a sad reflection of the current state of site search and evaluation in the UK that there is not a far greater use of the most advanced computer methods and remote sensing data are not being used. There is a complete almost off-the-shelf methodology for multicriteria evaluation using quantitative and qualitative data that are not being used. Surely the cost and importance of power developments can easily justify the use of

323

multi-million-pound site evaluation and automated site search methods. It is almost as if the CEGB conduct their business in the 1980s in virtually the same fashion as they did in the 1960s. If this is the case, then it is time there was a reorganisation of their management structures and techniques.

(6) Finally, British nuclear agencies would be well advised to be more willing to adopt latest US practices and to learn from the far more extensive US experience of nuclear power and especially of PWR operations. It seems that many knowledgeable US nuclear experts are amazed at the happenings in Europe and the immense naivety of both public and government. It would seem that the UK is perhaps 10 to 15 years behind the US with respect to safety and economic issues, for no apparently sound reasons.

This book has offered a largely unbiased appraisal of siting policies and practices in the UK and the US. It is emphasised that the locational decisions are of exceptional importance because of their likely longevity, the sums of public money which become fixed at particular sites, and the potential consequences of poor decisions for public health. Nuclear power seems to be exceptionally safe, but safety is something that can only be proved retrospectively by future generations. Current UK siting practices largely neglect the contribution to public safety that could be obtained by seeking less economic but potentially far safer sites. It has been argued that prudence and common sense would imply that as far as possible the geography of the UK should be exploited to minimise public risks from remote but not impossible reactor accidents. If any errors are to be made then they should err on the side of caution and public safety, rather than invoke spurious economic and environmental reasons to rule out the safest sites. It may be that the 1957 Electricity Act needs modification, if not a complete overhaul. Nevertheless, siting decisions should in the future take into account not just current patterns of consumption and current safety standards, but the need to anticipate the long-term situation when the safety standards may be quite different and demand patterns have changed. Sites being selected now will still be active in the twenty-second century, and this kind of consideration needs explicit attention. Failure to do so will amount to an unacceptable gamble with public safety, public funds, and the infrastructure needed for modern life.

Bibliography

Adams, C.A., Faux, F. (1969), 'Siting of CEGB nuclear power stations',
    Proceedings of Symposium on Safety and Siting, London, British
    Nuclear Energy Society, The Institute of Civil Engineers, pp. 37-43.
Adams, C.A., Stone, C.N. (1967), 'Safety and siting of nuclear power
    stations in the United Kingdom', in *Containment and Siting of
    Nuclear Power Plants*, Vienna, IAEA, 129-42.
Alderson, D.A., Bracewell, R.J., Rashid, J.A., Stevens, I. (1982), *The
    Siting of Nuclear Power Stations: A Case for Criticism*, Department of
    Town and Country Planning, Newcastle University, (mimeo).
Aldrich, D.C., Sprung, J.L., Alpert, D.J., Diegert, K., Ostmeyer,
    R.M., Richie, L.T., Strip, D.R., Johnson, J.D., Hansen, K.,
    Robinson, J. (1982), *Technical Guidance for Siting Criteria
    Development*, NUREG/CR 2239, SAND81-1549, Sandia National
    Laboratories, Albuquerque, New Mexico.
American Physics Society (1975), Report by the study group on light-
    water reactor safety, *Reviews of Modern Physics* 47, Supplement 1,
    1-124.
Andreas, I., Heide, R., Eberhard, J., Stolarz, H., Winken, R. (1979),
    *Kraftwerksstandorte aus der Sicht der Raumordung Schriftenreihe des
    Bundesministers fur Raumordung*, Bauwessen und Stadtebau, Bonn.
Arnold, P.J. (1982), 'System benefits of the PWR development at
    Sizewell and transmission connections', CEGB Proof of Evidence,
    Sizewell B Power Station Public Inquiry, CEGB, London.
Baker, J. (1981), *Nuclear Power – A Matter of Confidence*, London,
    CEGB.
Beattie, J.R. (1963), *An Assessment of Environmental Hazard from
    Fission Product Releases*, AHSB(S), R64, Harwell, United Kingdom
    Atomic Energy Authority.
Beattie, J.R. (1969), 'A review of hazards and some thoughts on safety
    and siting', *Proceedings of Symposium on Safety and Siting*, London,
    British Nuclear Energy Society, The Institute of Civil Engineers, pp.
    1-6.
Beckmann, P. (1976), *The Health Hazards of Not Going Nuclear*,
    Boulder, Colorado, Golem Press.

325

Bell, G.D., Charlesworth, F.R. (1962), 'Licensing and inspection of nuclear installations in the UK', in *Reactor Safety and Hazards Evaluation Techniques*, Volume 2, Vienna, IAEA, pp. 15-30.

Bell, G.D., Charlesworth, F.R. (1963), 'The evaluation of power reactor sites', in *Proceedings of Siting of Reactors and Nuclear Research Centres Conference*, Vienna, International Atomic Energy Agency, 317-29.

Briggs, R.B. (1978), *Feasibility of a Nuclear Siting Policy Based on the Expansion of Existing Sites*, ORAU/IEA-78-19(R), Institute for Energy Analysis, Oak Ridge, Tennessee.

Brown, G., Kronberger, H., Leslie, F.M., Moore, J., Mummery, P.W. (1958), 'Safety aspects of the Calder Hall reactor in theory and experiment', *Second United Nations International Conference on the Peaceful Uses of Atomic Energy*, New York, pp. 267-75.

Brown, M. (1976), *Health, safety and social issues of nuclear power and the nuclear initiative*, Stanford University, Institute for Energy Studies.

Brown, S. (1970), 'Cockcroft Memorial Lecture: The background to the nuclear development programme', *Journal of the British Nuclear Energy Society* 9, 4-10.

Bunch, D.F. (1978), *Metropolitan Siting: A Historical Perspective*, NUREG 0478, Washington DC, NRC.

Bunch, D.F., Bell, L.W., Farrell, C.M., Gibson, I.C., Murphy, K.G., Reyes, J.N., Soffer, L. (1979), *Demographic Statistics pertaining to Nuclear Power Reactor Sites*, NUREG 0348, Washington DC, NRC.

Burns, D. (1967), *The Political Economy of Nuclear Energy*, London, McCorquodale.

Burns, D. (1978), *Nuclear Power and the Energy Crisis*, London, Macmillan.

Burwell, C.C. (1976), *Nuclear Energy Center Site Survey – 1975 Part V, Section 4*, NUREG 0001, Washington DC, Nuclear Regulatory Commission.

Burwell, C.C., Lane, J.A. (1980), *Nuclear Site Planning to 2025*, ORAU/IEA-80-5A(M), Institute for Energy Analysis, Oak Ridge.

Burwell, C.C., Ohanian, M.J., Weinberg, A.M. (1979), 'A siting policy for an acceptable nuclear future', *Science* 204, 1043-51.

Caldicott, H. (1980), *Nuclear Madness*, Brookline Mass, Autumn Press.

Catchpole, S., Jenkins, F.P. (1977), 'UK experience of planning the nuclear contribution to the UK power programme', *Proceedings of Nuclear Power and its Fuel Cycle*, Volume 1, Vienna, IAEA, pp. 175-88.

Cave, L., Halliday, P. (1969), 'Safety and siting: A report of the BNES Symposium', *Journal of the British Nuclear Energy Society* 8, 180-84.

Cave, L., Halliday, P. (1969b), 'Suitability of gas cooled reactors for fully urban sites', *Proceedings of the British Nuclear Energy Society Symposium on Siting and Safety*, London, Institute of Civil Engineers, pp. 101-10.

Central Electricity Generating Board (1965), *An Appraisal of the Technical and Economic Aspects of Dungeness B Nuclear Power*

Station, London, CEGB.

Central Electricity Generating Board (1980), *Druridge: Site for a possible nuclear power station*, London, CEGB.

Central Electricity Generating Board (1980), *Electricity Supplies in the South-West: Power station site investigation*, London, CEGB.

Central Electricity Generating Board (1981), *Supplementary Information on Site Specific Aspects*, Sizewell B Power Station Public Inquiry, London, CEGB.

Central Electricity Generating Board (1982), *South-west Power Station Study*, London, CEGB.

Charlesworth, F.R., Gronow, W.S. (1967), 'A summary of experience in the practical application of siting policy in the United Kingdom', in *Containment and Siting of Nuclear Power Plants*, Vienna, IAEA, 143-70.

Charpentier, J.P. (1976), 'Towards a better understanding of the distribution of per capita energy consumption in the world', *Energy* 1, pp. 325-50.

Chenault, W.W., Hilbert, G.D., Reichlin, S.D. (1979), *Evacuation Planning in the TMI Accident*, Washington DC, Federal Emergency Planning Agency.

Chicken, J.C. (1981), *Nuclear Power Hazard Control Policy*, Oxford, Pergamon.

Chita, P.S. (1982), *Hartlepool Report*, Oxford, Political Ecology Research Group.

Clarke, R.H., Kelly, G.N. (1981), 'MARC – the NRPB methodology for assessing the radiological consequences of accidental releases of activity', *National Radiological Protection Board Report* 127, London, HMSO.

Coffin, B. (1984), *Nuclear Power Plants in the United States: Current Status and Statistical History*, New York, Union of Concerned Scientists.

Cope, D.F., Baumann, H.F. (1977), *Expansion Potential for Existing Nuclear Power Station Sites*, ORNL/TM-5927, Oak Ridge National Laboratory, Oak Ridge, Tennessee.

Cottrell, A. (1981), *How Safe is Nuclear Energy?*, London, Heinemann.

Countryside Commission (1968), *The Coasts of England and Wales: Measurement of Use, Protection, and Development*, London, HMSO.

Cutter, S. (1984), 'Residential proximity and cognition of risk at Three Mile Island: Implications for evacuation planning', in M.J. Pasqualetti and K.D. Pijawka, *Nuclear Power: Assessing and Managing Hazardous Technology*, Boulder, West View, pp. 247-58.

Cutter, S.L., Barnes, K. (1982), 'Evacuation behaviour and Three Mile Island', *Disasters* 6, 116-24.

Dale, G.C. (1982), *The Safety of the AGR*, London, CEGB.

Danckwerts, P. (1983), 'Windscale's pipeline', *New Scientist*, 15 December, p. 833.

Dell, E. (1973), *Political Responsibility and Industry*, London, Allen and Unwin.

Department of Energy (1979), *National Energy Policy*, Energy Paper No

41, London, HMSO.

Department of Trade and Industry (1972), 'Statement of Government siting policy for nuclear power stations', Connagh's Quay Public Inquiry, submission (mimeo).

Diggory, J.C. (1956), 'Some consequences of proximity to a disease threat', *Sociometry* 19, 47-53.

Dobson, J.E. (1979), 'A regional screening procedure for land use sitability analysis', *Geographical Review* 69, 224-34.

Dobson, J.E. (1983), 'Automated geography', *The Professional Geographer* 35, 135-43.

Drapkin, D.B. (1974), 'Development, electricity and power stations: problems in electricity planning decisions', *Journal of Public Law*, pp. 220-55.

Dunster, H.J. (1971), 'Environmental monitoring: British policy and procedures', in *Proceedings of Environmental Aspects of Nuclear Power Stations*, Vienna, IAEA, 427-37.

Dunster, H.J., Clarke, R.H. (1980), 'Remote siting brings small benefits', *Nuclear Engineering International* 25, 51-3.

Dunster, H.J., Latzko, D.G.H., Smidt, D., Villani, S. (1980), *Nuclear Safety in the Community of the European Communities: Report of the Expert Group on Nuclear Safety*, COM(80), 808, Brussels.

Durfee, R.C., Coleman, P.R. (1983), *Population Distribution Analyses for Nuclear Power Plant Siting*, NUREG/CR-3056, ORNL/CSD/TM-197, Oak Ridge National Laboratories, Oak Ridge.

El-Hinnawi, E.E. (1980), *Nuclear Energy and the Environment*, Oxford, Pergamon.

Electricity Council (1983), *Medium-Term Development Plan, 1983-90*, Electricity Council, London.

Farmer, F.R. (1962), 'The evaluation of power reactor sites', Proceedings of *Problemi di Sicurezza degli Impianti Nucleari: VII Congresso Nucleare*, Rome, CNEN, pp. 39-45.

Farmer, F.R. (1962b), 'The evaluation of power reactor sites', DPR/INF/266, United Kingdom Atomic Energy Authority, Harwell.

Farmer, F.R. (1967), 'Siting criteria – a new approach', in *Containment and Siting of Nuclear Power Plants*, Vienna, IAEA, pp. 303-18.

Farmer, F.R. (1979), 'A review of the development of safety philosophy', *Annals of Nuclear Energy* 6, 261-64.

Farmer, F.R., Fletcher, P.T. (1959), 'Siting in relation to normal reactor operation and accident conditions', *Symposium on Safety and Location of Nuclear Plant, Sixth International Electronic and Nuclear Congress*, Rome, CNEN, pp. 20-31.

Fernie, J., Openshaw, S. (1984), 'Policymaking and safety issues in the development of nuclear power in the United Kingdom', in M.J. Pasqualetti and K.D. Pijawka (eds), *Nuclear Power: Assessing and Managing Hazardous Technology*, Boulder, West View, pp. 67-92.

Flowers, B. (1976), *Nuclear Power and the Environment, Royal Commission on Environmental Pollution, Sixth Report*, London, HMSO.

Flynn, C. (1979), *Three Mile Island Telephone Survey: Preliminary*

Report on Procedures and Findings, NUREG/CR 1093, Washington DC, NRC.

Flynn, C. (1981), 'Local public opinion', in T. Moss, D. Sills (eds), *The Three Mile Island Nuclear Accident: Lessons and Implications*, New York, Annals of the Academy of Science, pp. 132-257.

Foley, G., van Buren, A. (1978), *Nuclear or Not: Choices for our Energy Future*, London, Heinemann.

Fry, T.M. (1955), *Population Distribution and Reactor Location*, RL/p 13, United Kingdom Atomic Energy Authority, Harwell, Oxford.

Fryer, D.R.H. (1969), 'Siting and safety evaluation in the UK', in *Proceedings of Symposium on Safety and Siting*, London, British Nuclear Energy Society, The Institute of Civil Engineers, pp. 47-60.

Gammon, K.M. (1979), 'CEGB experience in selecting and developing nuclear power station sites', *CEGB Newsletter* 111, London, Central Electricity Generating Board.

Gammon, K.M. (1981), 'The local impacts of nuclear stations', *Proceedings of Environmental Impact of Nuclear Power Symposium*, London, British Nuclear Energy Society, pp. 237-62.

Gammon, K.M. (1982), *Site Selection and Site Specific Aspects*, CEGB Proof of Evidence, Sizewell B Power Station Inquiry, P25, London, CEGB.

Gammon, K.M., Pedgrift, G.F. (1962), 'The selection and investigation of potential nuclear power station sites in Suffolk', *Journal of the British Nuclear Energy Society* 1, 220-34.

Gammon, K.M., Pedgrift, G.F. (1983), 'Changes in the investigation and selection of sites for nuclear power stations', *Nuclear Engineering* 22, 41-5.

Garner, J.F. (1970), *Administrative Law*, London, Butterworths.

Gatrell, A.C. (1983), *Distance and Space: a Geographical Perspective*, Oxford, Clarendon Press.

German Risk Study (1981), *A Study of the Risk due to Accidents in Nuclear Power Plants*, EPRI NP 1804-SR.

Golay, M.W. (1980), 'How Prometheus came to be bound: nuclear regulation in America', *Technology Review* 82, 29-39.

Gowing, M. (1964), *Britain and Atomic Energy, 1939-45*, London, Macmillan.

Gowing, M. (1974), *Independence and Deterrence – Britain and Atomic Energy, 1945-1952*, London, Macmillan.

Gronow, W.S. (1969), 'Application of safety and siting policy to nuclear plants in the United Kingdom', in *Environmental Contamination by Radioactive Materials*, Vienna, IAEA, pp. 549-59.

Gronow, W.S., Gausden, R. (1973), 'Licensing and regulatory control of thermal reactors in the United Kingdom', in *Principles and Standards of Reactor Safety*, Vienna, IAEA, pp. 521-38.

Hafele, W. (1977), 'Energy options open to mankind beyond the turn of the century', *Proceedings of Nuclear Power and its Fuel Cycle*, Vienna, Volume 1, pp. 58-81.

Hafele, W., Sassin, W. (1975), 'Application of nuclear power other than for electricity generation', *Research Report* 75-40, Luxemburg, IIASA.

Haire, T.P., Usher, E.F.F.W. (1975), 'Nuclear power station siting experience in the UK', in *Siting of Nuclear Facilities*, Vienna, IAEA, pp. 143-55.

Hallam, J., Hemming, C.R., Simmonds, J.R., Kelly, G.N. (1982), 'PROT-MARC the countermeasures module in the methodology for assessing the consequences of accidental releases', *National Radiological Protection Board* M 77, London, HMSO.

Hallam, J., Jones, J.A., Hemming, C.R. (1980), 'The establishment of population data grids for use in radiological protection studies', *National Radiological Protection Board* M54, London, HMSO.

Hamilton, D., Manne, A.S. (1977), 'Health and economic costs of alternative energy sources', in *Nuclear Power and its Fuel Cycle* Volume 7, Vienna, IAEA, pp. 196-214.

Hamilton, M.S. (1979), 'Power plant siting: a literature review', *National Resources Journal* 19, 75-95.

Haywood, S.M., Clarke, R.H. (1982), 'Degraded core accidents for the Sizewell PWR: a sensitivity analysis of the radiological consequences', National Radiological Protection Board Report 142, London, HMSO.

Health and Safety Executive (1979), *Safety Assessment Principles for Nuclear Power Reactors*, London, HMSO.

Health and Safety Executive (1981), 'The accident at Three Mile Island: comments by the Health and Safety Executive', in *First Report from the Select Committee on Energy 1980-1: Minutes of Evidence*, Volume 2, London, HMSO, pp. 19-23.

Health and Safety Executive (1982a), *The Work of the HM Nuclear Installations Inspectorate*, London, HMSO.

Health and Safety Executive (1982b), *Emergency Plans for Civil Nuclear Installations*, London, HMSO.

Hemming, C.R., Ferguson, L., Broomfield, M., Kelly, G.N. (1983), 'The consequences of accidental releases of radioactive material in extreme conditions', *National Radiological Protection Board* M79, London.

Hillsman, E.L., Alvic, D.R., Church, R.R. (1983), *BUILD (Baseload Utility Integrated Locational Decisions): a Model of the Future Spatial Distribution of Electric Power Production*, ORNL 5969, Oak Ridge.

Hinton, C. (1957), 'The future for nuclear power', *Royal Academy of Sciences*, Stockholm, 19-20.

Hinton, C. (1963), 'Nuclear power plants in England and Wales', *Atom* 59, pp. 165-9.

Hinton, C. (1967), 'Nuclear power development: some experiences of the first ten years', *Journal of the Institute of Fuel* 31, 90-108.

Hinton, C. (1969), 'Closing address', *Proceedings of Symposium on Safety and Siting*, London, British Nuclear Energy Society, The Institute of Civil Engineers, pp. 135-36.

Hinton, C. (1977), 'The birth of the breeder', in J.S. Forest (eds), *The Breeder Reactor*, Edinburgh, Academic Press, pp. 8-13.

Hobbs, B., Voelker, A. (1977), *Analytical Power Plant Siting Methodologies: A Theoretical Discussion and Survey of Current*

Practice, ORNL/TM-5749, Oak Ridge.

House of Commons Select Committee (1981), *Report on the Nuclear Power Programme*, London, HMSO.

Howells, G.D., Gammon, K.M. (1982), 'Role of research in meeting environmental assessment needs for power station siting', in R.D. Roberts and T.M. Roberts (eds), *Planning and Ecology*, London, Chapman and Hall, pp. 310-30.

Hunt, F.R. (1970), 'Power station site selection in England and Wales', in *Environmental Aspects of Nuclear Power Stations*, Vienna, IAEA, 647-59.

Hutber, F.W. (1984), 'Energy statistics and the planning perspective', *The Statistician* 1, 1-8.

Inhaber, H. (1978), *Risk of Energy Production*, Ottawa, Atomic Energy Control Board.

International Atomic Energy Agency (1978), *Safety in Nuclear Power Plant Siting*, Vienna, IAEA, 50-C-S.

International Atomic Energy Agency (1980), *Site Selection and Evaluation for Nuclear Power Plant with respect to Population Distribution*, Vienna, IAEA, 50-SG-S4.

International Atomic Energy Agency (1981), *External Man Induced Events in Relation to Nuclear Power Plant*, Safety Series No. 50-SG-S5, Vienna, IAEA.

Jeffery, J.W. (1982), 'A provisional economic critique of the CEGB's statement of case for a PWR at Sizewell', *Issues in the Sizewell B Inquiry*, Centre for Energy Studies, Polytechnic of the South Bank, London, Volume 3, pp. 20-35.

Johnson, J.H., Zeigler, D.J. (1984), 'A spatial analysis of evacuation intentions at the Shoreham nuclear power station', in M.J. Pasqualetti and K.D. Pijawka (eds), *Nuclear Power: Assessing and Managing Hazardous Technology*, Boulder, West View, pp. 279-302.

Joint Committee on Atomic Energy (1955), *Hearings on Development, Growth and the State of the Atomic Energy Industry*, Washington DC, US Government Printing Office.

Joint Committee on Atomic Energy (1971), *Hearing on AEC Licensing Procedures and Related Legislation*, Washington DC, US Government Printing Office.

Kay, K. (1956), *Calder Hall – the Story of Britain's First Atomic Power Station*, London, Methuen.

Keeney, R.L. (1980), *Siting Energy Facilities*, New York, Academic Press.

Kelly, G.N., Charles, D., Broomfield, M., Hemming, C.R. (1983), 'The radiological impact on the Greater London population of postulated accidental releases from the Sizewell PWR', *National Radiological Protection Board Report* 146, London, HMSO.

Kelly, G.N., Clarke, R.H. (1982), 'An assessment of the radiological consequences of releases from degraded core accidents for the Sizewell PWR', *National Radiological Protection Board Report* 137, London, HMSO.

Kelly, G.N., Hemming, C.R., Charles, D., Jones, J.A., Ferguson, L.,

Haywood, S.M. (1982), 'Degraded core accidents for the Sizewell PWR: a sensitivity analysis of the radiological consequences', NRPB R142, HMSO, London.

Kelly, M.J., Rush, R.M., Bauman, H.F., Ott, W.R. (1984), *Analysis of Availability of Previously Identified Sites under Alternative Demographic Criteria*, Oak Ridge National Laboratories, Oak Ridge.

Kemeny, J.G. (1980), *Report of the President's Commission on the Accident at Three Mile Island*, New York, Pergamon.

Kendrick, H., Cullen, C., McNeil, A., Vines, A. (1982), *Planning and hazardous industry*, Department of Town and Country Planning, Newcastle University (mimeo).

Kohler, J.E., Kenneke, A.P., Grimes, B.K. (1974), *The Site Population Factor: a Technique for Consideration of Population in Site Comparisons*, WASH 1235, Washington DC, NRC.

Lane, J.A., Covarrubias, A.J., Csik, B.J., Fattah, A., Woite, G. (1977), 'Nuclear power in developing countries', in *Proceedings of Nuclear Power and its Fuel Cycle*, Vienna, IAEA, Volume 1, pp. 231-49.

Lindsay, S., Norton, J. (1981), *Criteria for the Location of a Nuclear Power Station*, Department of Town and Country Planning, Newcastle.

Lundberg, U. (1973), 'Emotion and geographical phenomena', in R.M. Downs and D. Stea (eds), *Image and Environment*, Chicago, Aldine Press, pp. 322-37.

Macdonald, H.F., Ballard, P.J., Thompson, I.M.G. (1977), 'Recent developments in emergency monitoring procedures at CEGB nuclear power stations', in *The Handling of Radiation Accidents*, Vienna, IAEA, pp. 435-46.

Machta, L., Ferber, G.J., Heffter, J.L. (1974), 'Regional and global scale dispersion of 85kr for population dose calculations', in *Proceedings of Physical Behaviour of Radioactive Contaminants in the Atmosphere*, Vienna, IAEA, 411-26.

Marley, W.G., Fry, T.M. (1955), 'Radiological hazards from an escape of fission products and the implications in power reactor location', *Proceedings of the International Conference on the Peaceful Uses of Atomic Energy*, New York, UN, Volume 13, pp. 102-5.

Marshall, W. (1976), *An Assessment of the Integrity of the PWR Pressure Vessel*, United Kingdom Atomic Energy Authority, Harwell.

Marshall, W., Billington, D.E., Cameron, R.F., Curl, S.J. (1983), 'Big nuclear accidents', *Atomic Energy Research Establishment* Report 10532, London, HMSO.

Matthews, R. (1977), *Nuclear Power and Safety*, Central Electricity Generating Board, London.

Matthews, R.R. (1977b), *CEGB Experience of Public Communication*, Central Electricity Generating Board Newsletter 104, London.

Matthews, R.R. (1981), *Nuclear Power and Safety*, London, CEGB.

Matthews, R.R. (1982), *The CEGB Approach to Nuclear Safety*, Sizewell B Power Station Inquiry, CEGB Proof of Evidence, London, CEGB.

Matthews, R.R., Usher, E.F. (1977), 'CEGB experience of public communication', in *Proceedings of Nuclear Power and its Fuel Cycle*,

Vienna, IAEA, Volume 5, 145-51.

McCullough, C.R. (1955), 'The safety of nuclear reactors', in *Symposium on Safety and Location of Nuclear Plant, Sixth International Electronic and Nuclear Congress*, Rome, CNEN, Volume 13, p. 79.

McKinney, R. (1956), 'A forecast of the growth of nuclear fuelled electric generating capacity', in *Report of the Panel on the Impact of Peaceful Uses of Atomic Energy*, Washington DC, US Government Printing Office.

Meier, P. (1975), *Energy Facility Location: a Regional Viewpoint*, NTIS, BNL 20435, Brookhaven National Laboratories.

Mesler, R.B., Widdoes, L.C. (1954), 'Evaluating reactor hazards from airborne fission products', *Nucleonics* 12, 39-45.

Minogue, R.B., Eiss, A.L. (1976), 'Nuclear Regulatory Commission regulations and licensing', in J.H. Rust and L.E. Weaver (eds), *Nuclear Safety*, New York, Pergamon.

MITRE (1977), *Nuclear Power: Issues and Choices*, Cambridge, Massachusetts, Ballinger.

Montgomery, T.L., Rose, D.J. (1979), 'Some institutional problems of the US nuclear industry', *Technology Review* 81, 52-62.

Murphy, A.W., La Pierre, D.B., Orloff, N. (1978), *The Licensing of Power Plants in the United States*, Yale University, Seven Springs Center.

Niehaus, F. and Otway, H.J. (1977), 'The cost effectiveness of remote nuclear reactor sites', *Nuclear Technology* 34, pp. 387-97.

Nijkamp, P. (1980), *Environmental Policy Analysis*, London, Wiley.

Northumberland County Council (1979), *Nuclear Power Station Investigations: Druridge Bay*, Newcastle, Northumberland County Council.

Nuclear Power Company (1979), *Hartlepool Nuclear Power Station*, Whetstone, NPC.

Nuclear Regulatory Commission (1975), *Reactor Safety Study – an Assessment of Accident Risks in US Commercial Power Plant*, WASH 1400, NUREG 75-014, Washington DC, NRC.

Nuclear Regulatory Commission (1975b), *General Site Suitability Criteria for Nuclear Power Station, Regulatory Guide 4.7*, Washington DC, NRC.

Nuclear Regulatory Commission (1976), *Nuclear Energy Center Site Survey, 1975: Summary and Conclusions*, Washington DC, NRC.

Nuclear Regulatory Commission (1978), *Planning Basis for the Development of State and Local Government Radiological Emergency Response Plans in Support of Light Water Nuclear Power Plant*, NUREG 3096, EPA 520/1-78-016.

Nuclear Regulatory Commission (1979), *Report of the Siting Policy Task Force*, NUREG 0625, Washington DC, NRC.

Nuclear Regulatory Commission (1980), *Criteria for the preparation and evaluation of radiological emergency response plans and preparedness in support of nuclear power plants*, FEMA-REP-1 (Rev. 1), Washington DC, NRC.

333

Nuclear Regulatory Commission (1981), *Scoping Summary Report – Environmental Impact Statement on the Siting of Nuclear Power Plants*, NUREG 0833, Washington DC, NRC.

Nuclear Regulatory Commission (1978), *General considerations and issues of significance in the evaluation of alternative sites for nuclear generating stations under NEPA*, NUREG 0499, Washington DC, NRC.

Nuclear Regulatory Commission (1982), *Reactor Accident Source Terms: Design and Siting Perspectives*, NUREG 0773, Washington DC, NRC.

Office of Population Census and Survey (1980), *Mortality Statistics (cause) for England and Wales*, London, HMSO.

Office of Technology Assessment (1984), *Nuclear power in an Age of Uncertainty*, US Congress of Technology Assessment, OTA-E-216, Washington DC.

Olds, F.C. (1981), 'Emergency planning for nuclear plants', *Power Engineering* 85, 48-56.

Openshaw, S. (1980), 'A geographical appraisal of nuclear reactor sites', *Area* 12, 287-90.

Openshaw, S. (1982), *A Portable Suite of FORTRAN IV Programs (CCP) for Classifying Census Data for Districts and Counties: an Introduction and User Guide*, CURDS Report, Centre for Urban and Regional Development Studies, Newcastle University.

Openshaw, S. (1982a), 'The geography of reactor siting policies in the UK', *Transactions of the Institute of British Geographers*, New Series 7, 150-62.

Openshaw, S. (1982b), 'The siting of nuclear power stations and public safety in the UK', *Regional Studies* 16, 183-98.

Openshaw, S. (1983), 'Multivariate analysis of census data', in D. Rhind (ed.), *A Census User's Handbook*, London, Methuen, pp. 243-64.

Openshaw, S., Taylor, P. (1981), 'UK reactor siting policy: memorandum by the Political Ecology Research Group', in *First Report of the House of Commons Select Committee in Energy 1980-1: Minutes of Evidence* Volume 3, London, HMSO, pp. 990-1018.

Openshaw, S. (1984), 'An evaluation of the safety characteristics of current and possible future nuclear power station sites', *The Statistician* 33, 133-142.

Organization for Economic Co-operation and Development (1977), Energy production and environment', Paris.

Organization for Economic Co-operation and Development (1979), 'The siting of major energy facilities', Paris.

O'Riordan, T. (1984), 'The Sizewell B Inquiry and a national energy strategy', *The Geographical Journal* 150, 172-82.

Paley, W.S. (1952), *Resources for Freedom*, Washington DC, Presidential Commission on Material Policy, US Government Printing Office.

Parker, H.M., Healy, J.W. (1955), 'Environmental effects of a major reactor disaster', in *Proceedings of the International Conference on the Peaceful Uses of Atomic Energy*, Geneva, UN, Volume 13, pp.106-118.

Pask, V.A., Duckworth, J.C. (1955), 'The place of nuclear energy in the United Kingdom power development', *The Engineer* 200, 828-35.

Pasquill, F. (1961), 'The estimation of the dispersal of windborne material', *Meteorological Magazine* 90, 30-42.

Patterson, W.C. (1976), *Nuclear Power*, London, Penguin.

Penreath, R.J. (1980), *Nuclear Power, Man and the Environment*, London, Taylor and Francis.

Perry, R. (1977), 'Development and commercialisation of the light water reactor, 1946-1976', *Rand Corporation Report*, R-2180-NSF.

Pochin, E. (1976), *Estimated Population Exposure*, Paris, Nuclear Energy Agency, OECD.

Pocock, D.C.D., Hudson, R. (1978), *Images of the Urban Environment*, London, Macmillan.

Pocock, R.F. (1977), *Nuclear Power: Its Development in the United Kingdom*, London, Institute of Nuclear Engineers.

Policy Research Associates (1977), *Socioeconomic Impacts: Nuclear Power Station Siting*, NUREG-0150, Washington DC, NRC.

Political Ecology Research Group (1980), *Safety Aspects of the Advanced Gas Cooled Reactors*, Oxford, Political Ecology Research Group.

Power Station Impact Team (1982), *The Social and Economic Effects of Power Stations on their Localities*, Oxford, Oxford Polytechnic.

Preston, A. (1971), 'The UK approach to the application of ICRP standards to the controlled disposal of radioactive waste resulting from nuclear power programs', Vienna, IAEA, 147-57.

Putnam, P.C. (1954), *Energy in the Future*, New York, Macmillan.

Reid, G.L., Allen, K., Harris, D.J. (1973), *The Nationalised Fuel Industries*, London, Heinemann.

Rhind, D., Stanness, K., Evans, I.S. (1977), 'Population distribution in and around British Cities', *Census Research Unit* Working Paper 11, University of Durham.

Ritchie, L.T., Johnson, J.D., Blond, R.M. (1982), 'Calculation of reactor accident consequences, version 2 (CRAC2): computer code user's guide', *NUREG/CR-2326, SAND81-1994*, Albuquerque, New Mexico.

Roberts, P.C., Burwell, C.C. (1981), *The Learning Function in Nuclear Reactor Operation and its Implications for Siting Policy*, ORAU/IES 8-4(M), Institute for Energy Analysis, Oak Ridge, Tennessee.

Robinson, J.H., Hansen, K.L. (1981), *Impact of Demographic Siting Criteria and Environmental Suitability on Land Availability for Nuclear Reactor Siting*, Los Angeles, Dames and Moore.

Saunders, S. (1969), 'Opening address', *Proceedings of Symposium on Safety and Siting*, London, British Nuclear Energy Society, The Institute of Civil Engineers, pp. 1-3.

Schwarzer, W. (1968), 'The relationship between siting and safety of nuclear power stations', *Die Sicherheit standortunabhangiger Kernkraftwerke*, Paper 4, Dusseldorf.

Select Committee on Science and Technology (1969), *Report of the House of Commons Select Committee on Science and Technology in the Nuclear Power Industry, Minutes of Evidence and Appendices*, London, HMSO.

Select Committee (1967), *Report of the House of Commons Select Committee on Science and Technology on the UK Nuclear Reactor Programme: Report Minutes of Evidence and Appendices*, London, HMSO.

Select Committee (1976), *Report of the House of Commons Select Committee on Science and Technology on the SHGWR Programme*, London, HMSO.

Shaw, J., Palabrica, R.J. (1974), 'A critical review and comparison of the nuclear power plant siting policies in the UK and USA', *Annals of Nuclear Science Engineering* 1, 241-54.

Shrader-Frechette, J. (1980), *Nuclear Power and Public Policy*, New York, Reidel.

Sizewell B Public Inquiry (1983), *Transcript for Day 64*, (mimeo).

Sorenson, J.H. (1984), 'Evaluating the effectiveness of warning systems for nuclear power plant emergencies: criteria and application', in M.J. Pasqualetti and K.D. Pijawka (eds), *Nuclear Power: Assessing and Managing Hazardous Technology*, Boulder, West View, 259-78.

Starr, C., Whipple, C. (1982), 'Coping with nuclear power risks: the electric utilities incentive', *Nuclear Safety* 23, 1-7.

Stewart, N.G., Gale, H.J., Crooks, R.N. (1954), 'The atmospheric diffusion of gases discharged from the chimney of the Harwell pile (BEPO)', Atomic Energy Research Establishment, HP/R 1452.

Stoler, R. (1984), 'Pulling the nuclear plug', *Time* 7, 8-15.

Stott, M., Taylor, P. (1980), *The Nuclear Controversy*, Town and Country Planning Association, Oxford.

Strip, D.R. (1982), *Estimate of the Financial Consequences of Nuclear Power Reactor Accidents*, NUREG/CR 2723, SAND82-1110, Washington DC, NRC.

Suffolk Preservation Society (1982), *Dozens of Nuclear Power Sites: Highly Confidential List Published*, (mimeo).

Sweet, C. (1983), *The Price of Nuclear Power*, London, Heinemann.

Taylor, P. (1980), 'The implications for planners of an expanded nuclear programme', in Proceedings of PTRC Summer Conference, London, PTRC.

Toombs, F. (1977), *A review of nuclear power in the United Kingdom*, London, CEGB.

Union of Concerned Scientists (1977), *The Risks of Nuclear Power Reactors: a Review of the NRC Reactor Safety Study*, Cambridge Mass., Union of Concerned Scientists.

United Kingdom Atomic Energy Authority (1979), *The Development of Atomic Energy: Chronology of Events 1939-1978*, Harwell, HMSO.

United Nations (1955), *Peaceful Uses of Atomic Energy*, New York, UN.

Usher, R. (1962), 'The selection and investigation of potential nuclear power station sites in Suffolk', *Journal of the British Nuclear Energy Society* 1, 220-34.

US Atomic Energy Commission (1952), *Atomic Energy Commission*

Reactor Exclusion Distance Formula, Report by the Director of Reactor Development, AEC 172/13, Washington DC.

US Atomic Energy Commission (1957), 'Theoretical possibilities and consequences of major accidents in large nuclear power plant', *WASH* 740, Washington DC.

US Atomic Energy Commission (1959), *Notice of proposed rule making on site considerations*, FR-59-4342.

US Atomic Energy Commission (1960), *Report to Advisory Committee on Reactor Safeguards on the Operation of the PWR at increased Power Levels*, Washington DC, AEC.

US Atomic Energy Commission (1962), *Statement of Considerations, Part 100 Reactor Site Criteria*, 27/FR/3509, Washington DC.

US Atomic Energy Commission (1963), *Calculation of Distance Factors for Power and Test Reactor Sites*, TID-14844, Washington DC, ASAEC.

US Department of Energy (1982), *US Commercial Nuclear Power*, EIA-0315, Washington DC, DOE.

Vaughan, R.D., Chermanne, J. (1977), 'The need for high performance breeder reactors', in *Proceedings of Nuclear Power and its Fuel Cycle*, Volume 1, pp. 550-63, Vienna, IAEA.

Vaughan, R.D., Joss, J.O. (1964), 'The current and future development of the MAGNOX reactor', Proceedings of Anglo-Spanish Symposium, Madrid, Volume 1, Paper 4.

Voogd, H. (1983), *Multicriteria Evaluation for Urban and Regional Planning*, London, Pion.

Walmsley, D.J. (1978), 'Stimulus complexity in distance distortion', *Professional Geographer* 30, 14-19.

Weinberg, A.M. (1977), 'Nuclear energy at the turning point', in *Proceedings of Nuclear Power and its Fuel Cycle*, Vienna, IAEA, Volume 1, pp. 761-73.

Wilbanks, T.J. (1984), 'Scale and the acceptability of nuclear energy', in M.J. Pasqualetti and K.D. Pijawka (eds), *Nuclear Power: Assessing and Managing Hazardous Technology*, Boulder, West View.

Williams, R. (1980), *The Nuclear Power Decisions: British Policies 1953-78*, London, Croom Helm.

Wilson, A.G. (1970), *Entropy in Urban and Regional Modelling*, London, Pion.

Wilson, H. (1971), *The Labour Government 1964-1970*, London, Nicolson and Joseph.

Zeigler, D.J., Brunn, S.D., Johnson, J.H. (1981), 'Evacuation from a nuclear technology disaster', *Geographical Review* 71, 1-16.

Index